Environmental Division 2015

Core Programming Area at the 2015 AIChE Spring Meeting and 11th Global Congress on Process Safety

Austin, Texas, USA
26-30 April 2015

ISBN: 978-1-5108-0702-0

Printed from e-media with permission by:

Curran Associates, Inc.
57 Morehouse Lane
Red Hook, NY 12571

Some format issues inherent in the e-media version may also appear in this print version.

Printed by Curran Associates, Inc. (2015)

AIChE
3 Park Avenue
New York, NY 10016-5991

Phone: (203) 702-7660
Fax: (203) 775-5177

www.aiche.org

Additional copies of this publication are available from:

Curran Associates, Inc.
57 Morehouse Lane
Red Hook, NY 12571 USA
Phone: 845-758-0400
Fax: 845-758-2634
Email: curran@proceedings.com
Web: www.proceedings.com

TABLE OF CONTENTS

409505 Chemical Engineering: Is This the Golden Age?

Abstract:

Schlanger will offer his insights on the current opportunities in the chemical process industries in his talk entitled "Chemical Engineering: Is This the Golden Age?" on Monday morning, April 27. At the conclusion of his lecture, Schlanger will receive the AGILE Award in recognition of his leadership at CEVA Group, a $9 billion logistics supplier to industry worldwide; LyondellBasell, a major player in the plastics, chemicals, and refining industries; Arco Chemical; and at the other companies where he has held leadership positions.

400449 Effect of Ozone Concentrations from Local Emission Sources

Abstract:

Ground level ozone is one of six criteria air pollutants designated by US EPA. It is not emitted directly into the air, but is created by chemical reactions between oxides of nitrogen (NOx) and volatile organic compounds (VOC) in the presence of sunlight. Emissions from industrial facilities and electric utilities, motor vehicle exhaust, gasoline vapors, and chemical solvents are some of the major sources of NOx and VOC.

In this report, rigorous air quality simulation was performed to simulate the ozone formation of the Beaumont & Port Arthur, Texas (BPA) area. Simulation results match observed data from monitoring stations. Ozone source apportionment showed local emission sources contributed 20% of total ozone. For the local contributions, point souces such as industrial facilities and electric utilities contributed 10% of total ozone, and on-road vehicles contributed 5%. Matrix simulations and EKMA diagrams were also performed to show the effects of emission reductions of anthropogenic NOx and VOC sources.

Emissions Testing of Sonic Velocity Flares Validates High Destruction and Removal Efficiency

Scot Smith
Director of Flare Products
Zeeco, Inc.
22151 East 91st Street
Broken Arrow, Okla. 74014
Scot_smith@zeeco.com

ID #400369

Prepared for Presentation at
American Institute of Chemical Engineers
2015 Spring Meeting
11th Global Congress on Process Safety
Austin, Texas
April 27-30, 2015

UNPUBLISHED

Abstract

In a 2012 report by the EPA's Flare Review Panel, regulations for flares that tended to operate outside their stable flame envelope were expanded to include limits on velocity as a function of net heating value.

EPA regulations do not currently allow sonic velocity flares to be permitted and operated without first performing an alternate means of emissions limitation test to validate destruction efficiencies. Existing regulations affect the application of multi-point staged flares since these flares operate at sonic exit velocities that exceed the maximum exit velocity requirements cited in 40 CFR 60.18 and 40 CFR 63.11(b) and applicable state regulations.

Over the years, Zeeco has conducted numerous destruction efficiency tests on its multi-point sonic flares. All of these tests demonstrated high destruction efficiency at velocities exceeding the EPA's limits. The purpose of these tests, performed at Zeeco's industrial-scale combustion test facility near Tulsa, Oklahoma, was to validate sonic tips for more applications using an expanded range of gases and conditions. Extractive sampling was used to measure the emissions from the exhaust plume at staging and de-staging pressures. The extracted gas concentrations were used to calculate the destruction efficiency of the multi-point sonic flare tips at each test point. Even with sonic gas exit velocities, the Zeeco-designed sonic flare tips maintained high destruction efficiency.

Despite pressure from both end users and regulators, Zeeco's flare burner tests and results have been maintained internally. This paper will present those results to the industry for the first time.

Benefits and Characteristics of Multi-Point Ground Flares

Multi-Point Ground Flares (MPGF) derive their name from their physical layout. Typically, they are a field of multiple pressure-assisted flare tips that are vertically mounted at grade and arranged in stages that open as the upstream pressure (gas flow) increases, and close as pressure decreases. This design concept is the same today as it was when MPGF were invented over 40 years ago. MPGF are often selected to attain smokeless combustion for heavy hydrocarbon service with high available pressure, and have proven to provide stable operation at sonic velocities for a wide range of gas compositions. MPGF are also used in situations where the designer wants to reduce or eliminate radiation or visible flame. The high pressure is used to assist the gas to obtain smokeless operation for the full range of flaring capacity, which can be difficult to do with other assist mediums such as air, gas, or steam due to high smokeless rates. Each burner has unobstructed air access allowing the momentum from the high exit velocity of the flare gas to entrain the necessary air for combustion. MPGF are designed to provide maximum smokeless performance, while minimizing radiation impacts and plot space. Installing a fence around the field can block the visibility of the flame; this serves a dual purpose by reducing radiation outside the fenced area and reducing the likelihood that flaring operations will be a nuisance to the public. The figure below is an example of a typical MPGF installation.

Figure 1 – Typical Installation of a MPGF

Another benefit of MPGF is easy access for maintenance, because all of the staging equipment is located at grade and outside of the fence. As a result, personnel can access the staging equipment safely without being affected by a flaring event. Figure 2 illustrates the staging equipment and the exterior of a MPGF fence.

Figure 2 – Typical Installation of MPGF Fence

Over the past 40-plus years, there have been many improvements in burner technology. Zeeco's modern pressure-assisted tips are custom designed to optimize the mixing of flare gas and air thus improving tip performance. Zeeco's pressure-assisted flare tips, shown in Figure 3, are an investment casting typically made out of 310 SS casting material.

Figure 3 – Zeeco's Pressure-Assisted Flare Tip Casting

Multiple tips are arranged in stages to allow control over the number of tips in service depending upon the pressure and flow rate of the flare gas. This enables the tips to operate in the optimal pressure range for maximum smokeless capacity. Each stage will open when the pressure in the MPGF header has increased to the designed staging pressure and subsequently close when the de-stage pressure is met. Figure 4 is an example of a typical staging curve used to control MPGF.

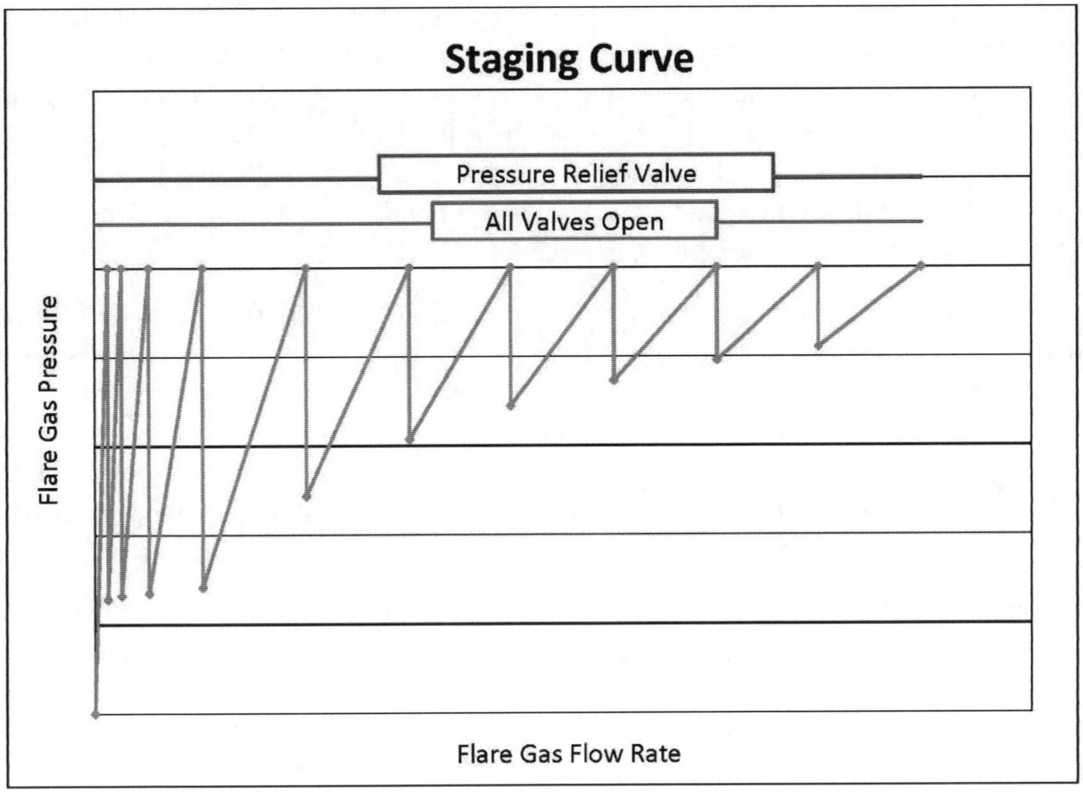

Figure 4 – Typical Staging Curve for MPGF

The staggered blue line in Figure 4 represents how the pressure in the system will change as each stage opens and closes. The high and low points of the blue line represents the staging and de-staging pressures for each stage, respectively. Typically, the staging pressure is equal to the available pressure of the system; however, it could be lower based on maintaining stable performance. When the pressure in the MPGF header reaches the designed staging pressure, the next stage of tips opens. The system is designed so that the pressure in the open stages never decreases below a desired pressure. This is referred to as the de-staging pressure and generally corresponds to the minimum pressure required to ensure stable performance and smokeless combustion. A typical de-staging curve follows the same pattern as the staging curve but closes the valves as the pressure in the MPGF header decreases.

Regulations and Industry Testing

Currently, federal and state regulations on flares limit the exit velocity of the flare gas based on the composition and LHV of the gas as well as the assist medium of the flare. There are three assist types considered by the regulations: non-assisted, steam-assisted, and air-assisted. According to the federal and state regulations, if a flare is operated within these velocity and LHV restrictions, a destruction and removal efficiency (DRE) of 98% or greater is guaranteed. This suggests that if a flare operates outside of these limits, it will result in lower DRE.

In 1983, flare efficiency testing was performed by the Chemical Manufacturers Association (CMA). Three types of flares were tested: air-assisted, steam-assisted, and non-assisted. It is important to note that pressure-assisted flares were not included in this testing. In the testing performed, extractive sampling was used to measure the concentrations of emissions from each flare to determine the combustion efficiency (CE). The results from this testing became the basis of the current regulations: 40 CFR 60.18 & 40 CFR 63.11. The CMA concluded if there was a stable flame, the flare had high CE.

From 1984-86, Energy and Environmental Research (EER) Corporation performed testing for the EPA to evaluate the combustion efficiency CE for a variety of gas compositions on different commercially available flare tips, including two pressure-assisted tips that were obtained from flare vendors. The pressure-assisted flare tips were referred to in the test report as Commercial Tips "E" and "F." Commercial Tip E used a horizontal bar geometry while Commercial Tip F used an open geometry. The results showed that stable burning pressure-assisted tips operated at CE values of greater than 98%, even at high exit velocities.

In November 2013, Dow Chemical Company performed an Alternate Means of Emissions Limitation (AMEL) test for two of its MPGF installations. Nominal four-inch sonic burners were tested by extractive sampling to investigate the propylene DRE of two different flare tips, one of which was pressure-assisted. The results from this testing showed that the pressure-assisted tips were capable of providing greater than 98% DRE and CE over the range of gases that were tested. Furthermore, the testing confirmed that the presence of a stable flame ensured high DRE and CE. The conclusions and results were published and presented at the American Flame Research Committee (AFRC) Conference in 2014.

The current EPA regulations are based on the CMA testing in 1983; however, various other tests have been performed to further investigate pressure-assisted flare efficiency. These tests prove sonic flares can operate outside of the limits set by the EPA and still achieve high DRE.

Execution of Extractive Sampling

In addition to the testing previously described, Zeeco has conducted numerous destruction and removal efficiency tests on its MPGF pressure-assisted tips. All MPGF testing that is further detailed in this paper was conducted at the Zeeco Inc. Combustion and Research Test Facility in Broken Arrow, Oklahoma. The purpose of Zeeco's tests on MPGF was to investigate the destruction and removal efficiency and stability over a wide range of net heating values (NHV) and gas compositions. NHVs from 440 BTU/SCF to 2316 BTU/SCF were tested with gas mixtures including the following: propane, propylene, natural gas, carbon dioxide, nitrogen, and hydrogen. Destruction and removal efficiency was determined by measuring the concentration of carbon dioxide (CO_2), carbon monoxide (CO), and total hydrocarbons (THC) from extractive sampling. An example of extractive sampling being performed on a single MPGF pressure- assisted tip is shown in Figure 5.

Figure 5 – Extractive Sampling Method Used for Testing

Zeeco's custom sample hood, shown in Figure 5, was used to extract combustion products from the plume. A Venturi nozzle located at the end of the apparatus created a vacuum for sample extraction. A crane was used to position the inlet of the sample hood in the plume of combustion products. The crane was directed by an operator who monitored the hood's position with respect to the plume using a forward looking infrared (FLIR) camera to ensure a proper sample was obtained. A continuous sample was pulled from the sample hood through a heated sample line to gas analyzers where the concentration of the combustion products was measured and recorded.

The concentrations of CO_2, CO, and THC were used to calculate destruction and removal and combustion efficiency. DRE and CE are two ways of quantifying the degree of completion of the combustion reaction based on measured emissions. DRE is how well a component of interest is destroyed or broken down according to the amount of unburned hydrocarbons after the combustion process is completed. Alternatively, CE is how well a component of interest is converted into CO_2 and H_2O after the combustion process. The hydrocarbon DRE and CE equations are shown below.

$$DRE_{THC}(\%) = \frac{CO_2 + CO}{CO_2 + CO + THC} \cdot 100\%$$

$$CE(\%) = \frac{CO_2}{CO_2 + CO + THC} \cdot 100\%$$

It is important to understand the difference between both values to properly interpret the results. The CE is equal to or less than the DRE since the component of interest may be destroyed, but not completely combusted. The component of interest can reduce to an intermediate combustion product instead of completely combusting and forming only CO_2 and H_2O. To summarize, generating proportionally high CO_2 will result in high CE, while generating proportionally low hydrocarbons will result in high DRE. For conservative results, the DRE and

CE of total hydrocarbons were observed to verify that all of the hydrocarbons were being combusted.

Results

The following results are from numerous tests performed at Zeeco from 2013 to 2015. Gas mixtures with NHVs ranging from 440 to 2316 BTU/SCF were tested on a single piloted sonic flare tip to investigate the effect of NHV on flame stability and efficiency. The maximum allowable exit velocity per 40 CFR 60.18 is based on the NHV of the flare gas. For NHVs ranging from 200 BTU/SCF to 1000 BTU/SCF, an exponential function limits the exit velocity from 45.7 ft/s to 400 ft/s. For NHVs greater than 1000 BTU/SCF, the maximum allowable exit velocity remains 400 ft/s. Every gas was tested at exit velocities that exceeded the limits defined by 40 CFR 60.18. As long as the flame was stable, DRE and CE values greater than 99% were consistently observed. Figure 6 illustrates the high destruction and removal efficiencies obtained outside of the 40 CFR 60.18 operating limitations.

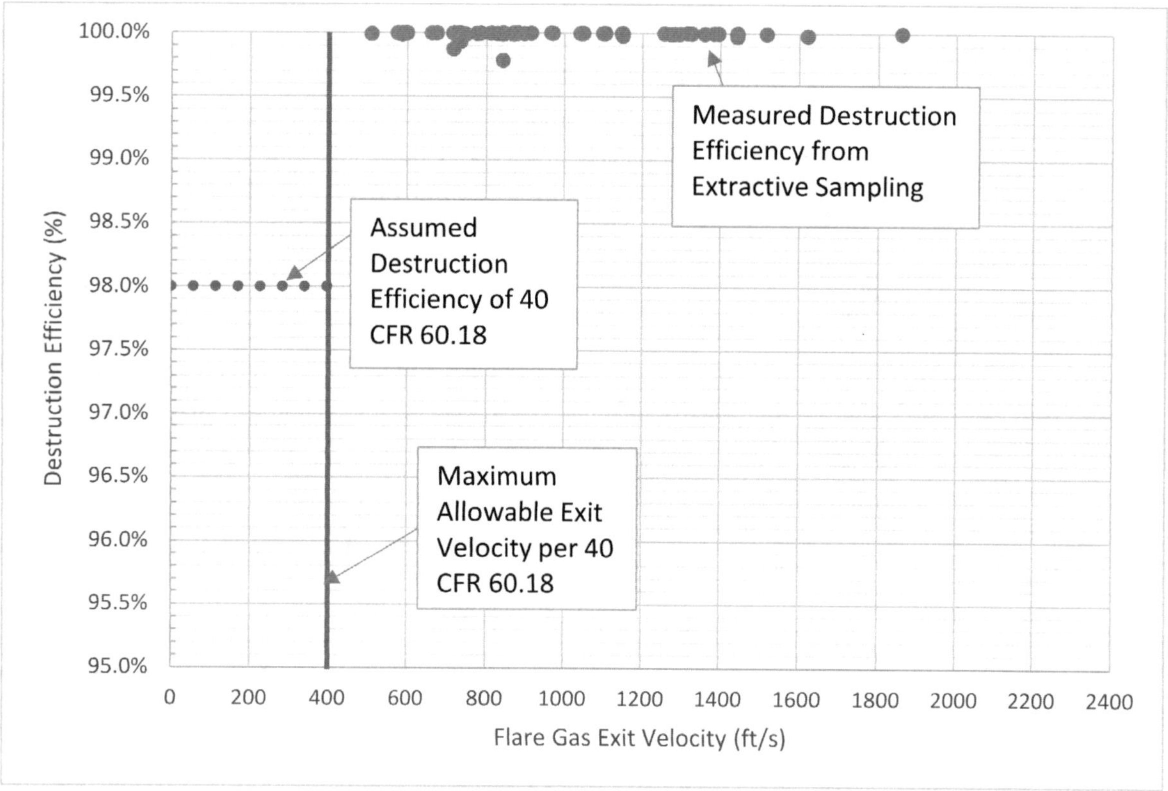

Figure 6 – Destruction and Removal Efficiency vs. Flare Gas Exit Velocity

The information in Figure 6 represents a total of 64 test runs. Flare gas mixtures ranged from molecular weight (MW) of 6.58 to 44.1 and from NHVs 440 to 2316 BTU/SCF. The mixtures were tested at operating pressures from 3 to 30 psig. The total compilation of all data was minimized by omitting the bottom 6% of the DRE values. The omitted test points will be discussed later.

Every gas combination tested that maintained a stable flame operated with high efficiencies despite the flare gas exit velocity exceeding the limits set forth in 40 CFR 60.18. The results from these tests on Zeeco's pressure-assisted tips prove that pressure-assisted flares should not have these exit velocity restrictions.

Several tests were performed to investigate the stability and effect of hydrogen (H_2) in hydrocarbon gas mixtures. The NHVs of the mixtures ranged from 440 to 1076 BTU/SCF with hydrogen mole percents between 25% and 70%. All gas mixtures were tested at exit velocities between 0.6 Mach (777 ft/s) and 1.0 Mach (2408 ft/s). Throughout the testing with H_2, the flame remained stable and attached to the tip and resulted in DRE and CE values of 99% or higher.

According to 40-CFR 60.18, an alternate method of calculating maximum exit velocity can be used for flare gas with a H_2 volume percent of 8.0 or greater. This alternate method linearly varies maximum exit velocity based on H_2 content between 8.0 percent at 25.6 ft/s and 15.6 percent at 122 ft/s and limits all volume percents greater than 15.6 to 122 ft/s. All of the H_2 mixtures tested had greater than 15.6 volume percent of H_2 and per 40 CFR 60.18 would be required to have an exit velocity less than 122 ft/s to meet the current EPA regulations. The results from Zeeco testing show that exit velocities greater than 122 ft/s did not negatively impact the flare efficiency for the H_2 mixtures tested. Although adding H_2 typically lowers the NHV of a gas mixture, testing showed that the presence of H_2 improved the stability of the flame compared to similar NHV mixtures without H_2. The benefit of H_2 flammability outweighed the downside of the reduced NHV, yielding high efficiencies.

The only unstable flame observed was when burning a gas mixture of propylene and nitrogen with an NHV of 600 BTU/SCF. This gas case was unable to maintain stability at the pressures tested and exhibited substantially lower DRE and CE than all other cases. In comparison, a mixture of natural gas and nitrogen with a NHV of 600 BTU/SCF remained stable at the pressures tested and maintained DRE and CE values above 99%. This suggests efficiency is not solely dependent upon the NHV of the flare gas. The difference in efficiency could be attributed to the volumetric ratio of flammable gas versus inert gas. In order to bring the NHV of a propylene mixture down to 600 BTU/SCF, a higher volume of inert gas is needed compared to a natural gas mixture. For example, the propylene mixture had approximately 73% volume inert gas where the natural gas mixtures had approximately 35% volume of inert gas. The higher volume of inert gas could be the cause of the lower efficiency.

Additional testing performed by Zeeco included the use of Passive Fourier Transform Infrared (PFTIR) in parallel with extractive sampling to measure CE. Four tests of propane, propylene, natural gas, and a propane and nitrogen mixture were used for this comparison. A similar exercise was conducted for Providence Engineering's new technology, FlareSENTRY.

The results from FlareSENTRY were extremely close to the results from the extractive method. All of the methods showed all four gas compositions had high CE while operating at sonic exit velocities as shown in Figure 7.

Gases	C3H8	C3H8/N2	C3H6	NG
NHV (BTU/SCF)	2316	1251	2183	937
40 CFR Maximum Allowable (ft/s)	400	400	400	400
Exit Velocity (ft/s)	841.4	969.9	869.8	1443.5
Mach Number	1.00	1.00	1.00	1.00
Flare Operating Pressure (psig)	16.0	10.3	16.9	15
CE (%) from Extractive Sampling	99.99%	99.99%	99.96%	99.99%
CE (%) from PFTIR	99.60%	99.90%	99.60%	99.50%
DRE (%) from Extractive Sampling	99.99%	99.99%	99.99%	99.99%
DRE (%) from FlareSENTRY	99.80%	99.55%	99.90%	99.70%

Figure 7 – Comparison of PFTIR vs. Extractive Sampling Method

FlareSENTRY is an optical-based monitoring technology developed by Providence Photonics, LLC, which remotely measures the DRE of flares. The unique capabilities of this new technology allow for continuous and autonomous monitoring of DRE with no operator input required. FlareSENTRY can also monitor the pilot ignition and provide a smoke index that can be used to predict the presence of smoke in the combustion plume. This technology will provide flare operators with real-time data to ensure optimal DRE while maintaining smokeless performance day or night. An industrial interface allows for integration with Distributed Control Systems (DCS) to enable closed loop operations. FlareSENTRY is a revolutionary new product that will redefine emissions monitoring and measurement for flares.

Conclusion

Multi-Point Ground Flares deliver smokeless flaring over a wide range of flows, compositions, and pressures while having minimal impact on surrounding communities. The testing conducted at the Zeeco Combustion and Research Test Facility investigated the effect of NHV and exit velocity on the stability and efficiency of Zeeco's pressure-assisted flare tips. All test results with stable flames demonstrate Zeeco's pressure-assisted flare tips provide high destruction and removal efficiency even at high exit velocities exceeding current regulatory limits. The extractive sampling results validated the performance of the new FlareSENTRY which is pioneering a practical method of monitoring and measuring flare efficiency. This testing has reinforced the industry consensus that pressure-assisted flare tips are capable of providing smokeless flaring with high destruction and removal and combustion efficiencies over a wide range of gas compositions and pressures. Recommendations for future testing are to investigate the stability of lower NHVs over a range of pressures and exit velocities as well as the effect of different ratios of combustible to inert gases. Current and future regulations should consider all available test results and allow the use of pressure-assisted flares with high exit velocities, up to and including Mach 1.0, without the need to perform an AMEL.

401845 Chloride Management in Naphtha Reforming with Continuous Catalyst Regeneration: A Method to Reduce Chloride Emissions and Consumption

Abstract:

Naphtha reforming with continuous catalyst regeneration (CCR) is an important process for the production of high octane gasoline and aromatics. Naphtha reforming catalysts are bi-functional catalysts that contain both acid and metal functions. The acid function is provided by chloride on the alumina base; as a result, chlorine containing compounds are present in the regenerator vent gas. In order to comply with environmental regulations many refiners must treat the regenerator vent gas to remove chlorine containing compounds. One method for reducing chloride emissions is the use of a conventional vent gas caustic scrubbing system which consumes caustic and generates hazardous waste. This contribution will compare a conventional vent gas scrubbing system to a system that captures chloride from the vent gas and recycles it back into the process, resulting in a 70% decrease in chloride consumption while simultaneously eliminating caustic usage.

399537 Sustainability Key Performance Indicators

Abstract:

Sustainability and climatle action require data and metrics to successfully measure performance towards goals and targets. The City of Austin has been pursuing climate action goals since 2007 and city-wide sustainabiltiy metrics since 2010. This presentation will explore sustainability and climate metrics, data collection, analysis, management, reporting, and successfully using this information to drive performance and change. With 4-5 years of data from city operations, which include 250 buildigs, 12,000 employees, 5,000 vehicles, a municipally owned water and power utility, as well as community wide sustainability outcomes, the amount of data and opportunities to communicate data are immense.

400440 Making Sustainability Tangible: Resilience, Behavior Change, and Risk

Abstract:

When corporations push forward on projects based on cost-benefit analyses that weigh social, economic, and environmental factors; the economic element of the triple bottom line often outweighs the other two components. Why is this? The answer is…..substance. When it comes to sustainability, managers respond most effectively when the outcomes move beyond the qualitative to the quantitative. In addition, sustainability has far reaching impacts on management decisions when approached through the lens of resilience and externalities. This presentation will explore scenarios and pose key questions about the critical importance of not only heavily incorporating all three components of the sustainability pyramid, but also how enhancing the social and environmental facets can cause unexpected paradigm shifts. Case studies will include disaster recovery and resilience, green office/operations practices, and employee engagement.

398145 How Sustainability Is Infiltrating Business Planning

Abstract:

The executive responsible for sustainability within an organization often serves in the role of management consultant, coach and advocate for sustainability practices throughout the organization. While many may find their influence limited to responding to sustainability questionnaires, others are gaining traction infiltrating sustainability into the myriad aspects of business planning. In many cases the bright line between a sustainabilit-driven activity, and general business practice is dimming. Drawing on case studies this talk will examine examples where sustainability practices are mainstreaming...and why. Examples will include corporate activities around water stewardship, product innovation, and supply chain management.

Improving Sustainability:

Employees Can Design/Build the Car, but Management Holds the Car Keys

Paul Westbrook
Sustainable Development Manager, LEED AP
Texas Instruments

April 2015

TEXAS INSTRUMENTS

Outline

- Sustainability Defined
- Technology vs Technique (Process)
- Integrated Design
- TI Case Study
- TI Results

- Abstract: The adoption of more sustainable practices hinges on two key areas. One is the technology development that allows for improved processes. The other, which is more difficult, requires changes in our behavior, habits, and even our culture. Management sets the expectations and examples for the company culture and can have a tremendous influence on the speed and depth of sustainable practice adoption. Texas Instruments (TI) has implemented a variety of sustainability initiatives over the years. This presentation will explore the key factors that led to successful initiatives and review the results of those initiatives. Some of TI's successes have been building the first LEED Gold semiconductor manufacturing facility in the world and cutting chip production energy intensity in half in just over 8 years.

TEXAS INSTRUMENTS

Sustainability Defined

- Official Definition

Brundtland Commission of the United Nations, 1987

Simply stated . . .
The balance of people, profit, and the planet

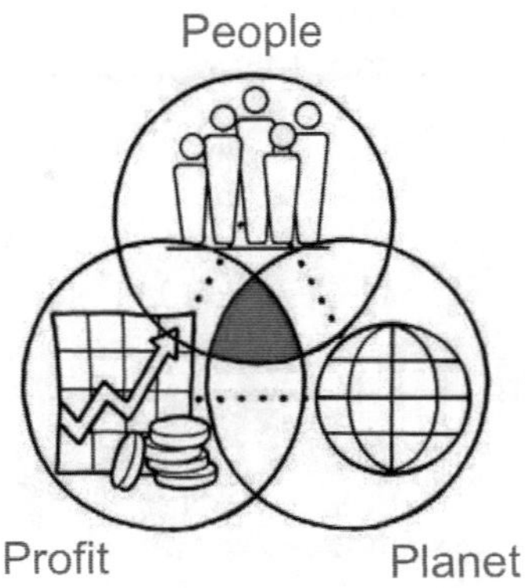

Texas Instruments

Technology vs Technique

Technology

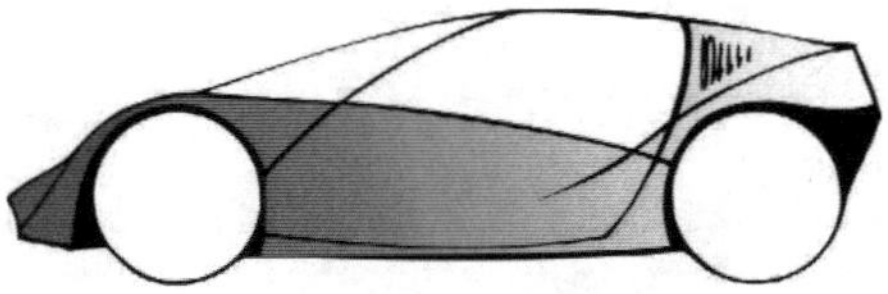

Machinery and equipment developed from the application of scientific knowledge

Technique (or Process)

A way of carrying out a particular task
How we organize our work flow/teams

Texas Instruments

Technology vs Technique

Technology

- LED lights
- VFD motors
- Efficient / optimized systems
- Energy dashboards

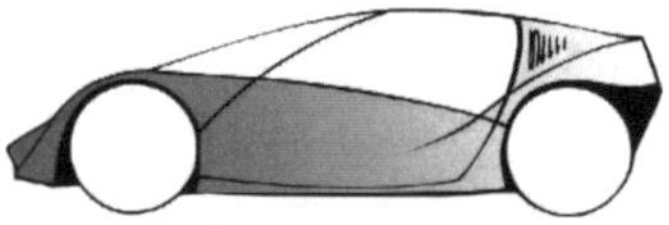

Technique

- Behavior
- Habits
- Culture
- Ethics
- Trust
- Support
- Organization

TEXAS INSTRUMENTS

Management Style Spectrum

Servant Leadership ⟷ **Top Down Micromanagement**

Servant Leadership	Top Down Micromanagement
Asking / Listening	Telling
People oriented / Empathy	Task-oriented
Awareness	Controlling
Persuasion	Directing
Planning / Foresight	Deadline driven
Stewardship	Done at all costs
Motivating	Driver
Creative	Risk-averse
Constructive	Authoritarian
Focus on people's strengths	Focus on "fixing" people's weaknesses

TEXAS INSTRUMENTS

The Design Process

"If it ain't broke, don't fix it."

Note: It's broken

TEXAS INSTRUMENTS

Integrated Design – Building Example

An iterative, non-linear process

Typical Linear Design Process

1. Architect draws up design

2. Everyone else makes it work:
- High cooling load window locations
- No room for ductwork
- Excessive materials use / waste
- Poor daylighting

. . . Over budget,
then "value engineering" occurs

Integrated Design Process

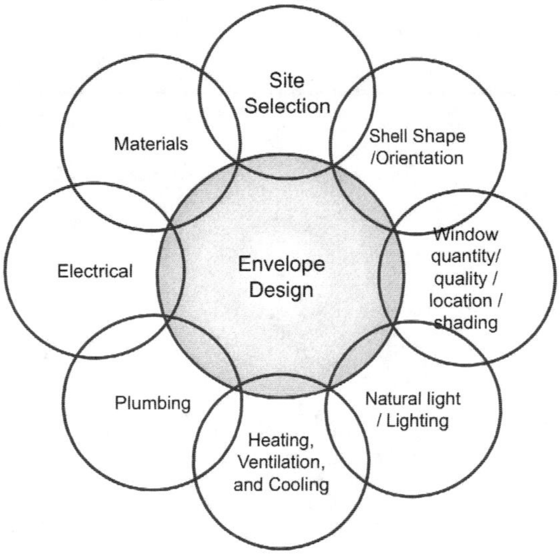

TEXAS INSTRUMENTS

Integrated Design

Value Engineering is Neither

- Amory Lovins, RMI

TEXAS INSTRUMENTS

Integrated Design: Loops

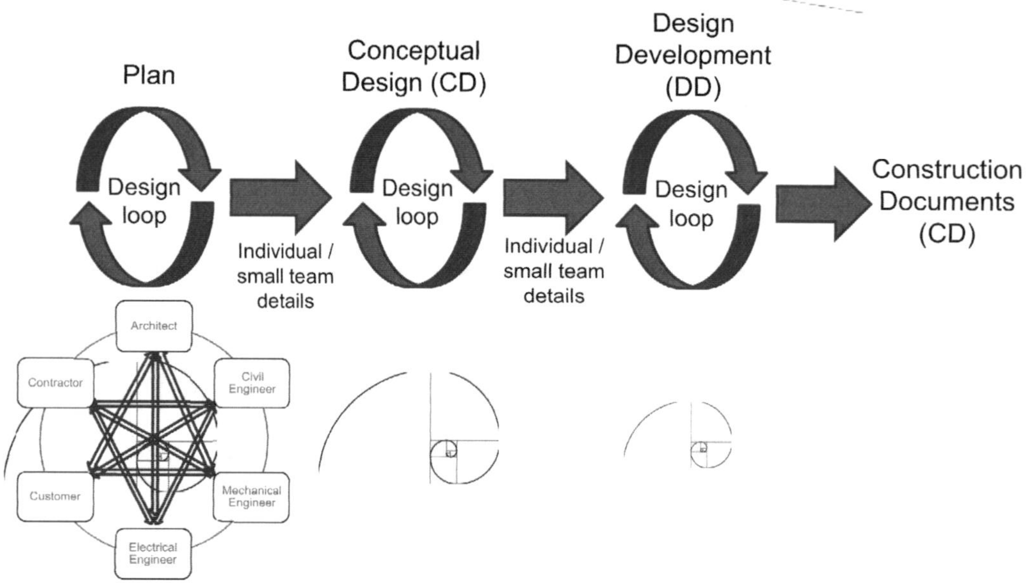

TEXAS INSTRUMENTS

Texas Instruments Example

TEXAS INSTRUMENTS

TI Business

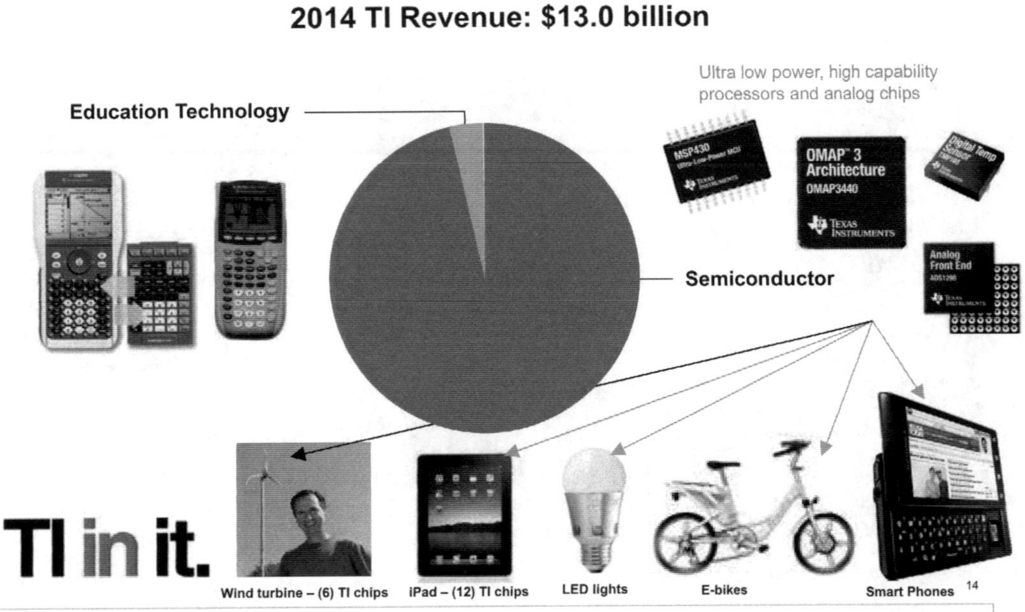

TEXAS INSTRUMENTS

TI's Global Presence

Approximately 32,000 employees worldwide

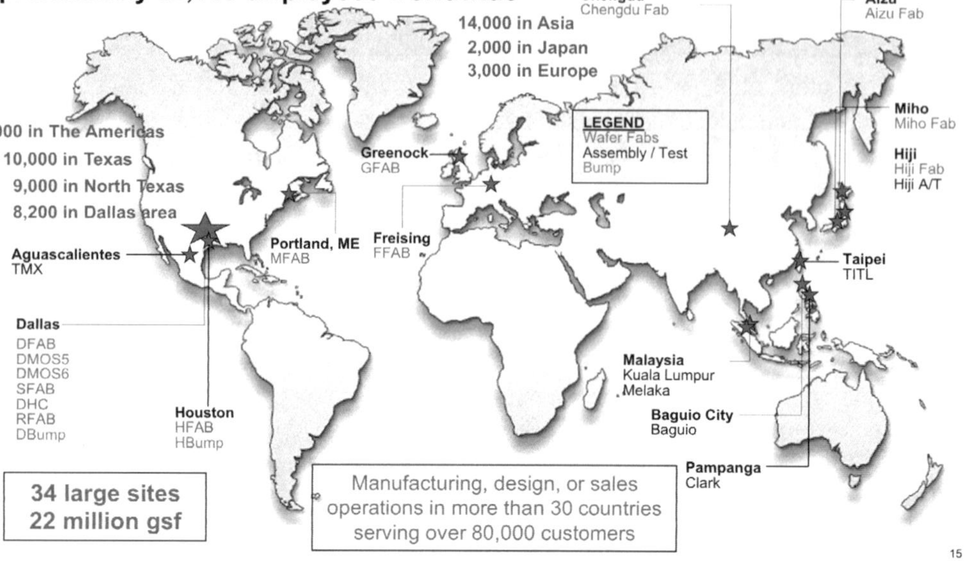

34 large sites
22 million gsf

Manufacturing, design, or sales operations in more than 30 countries serving over 80,000 customers

15

TEXAS INSTRUMENTS

What is a wafer fab?

A very big, clean facility . . .

Total space:
1,100,000 gsf (102,000 m2)
Clean room space:
220,000 sf (20,400 m2)

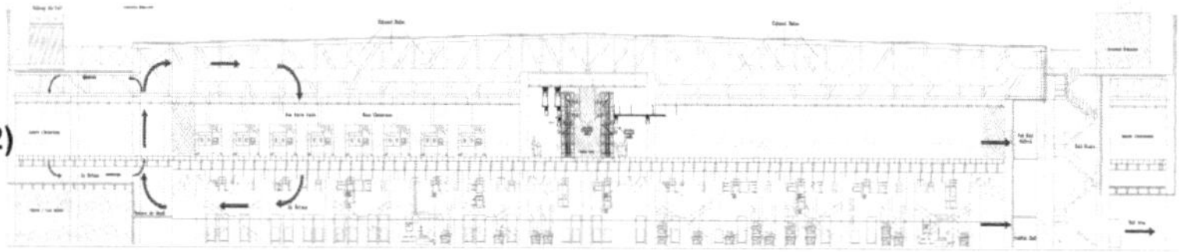

. . . that fabricates very small chips on large silicon wafers

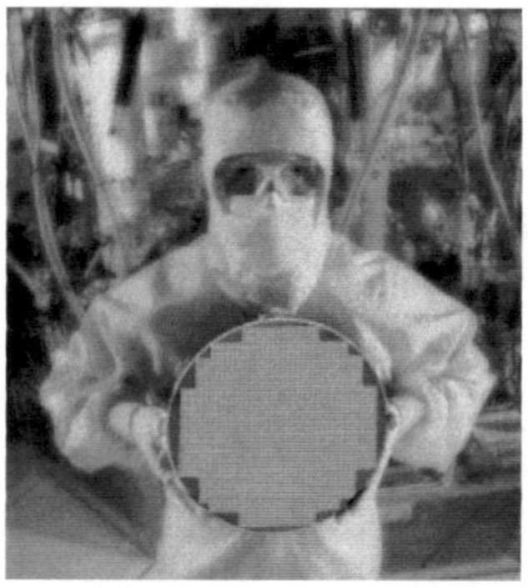

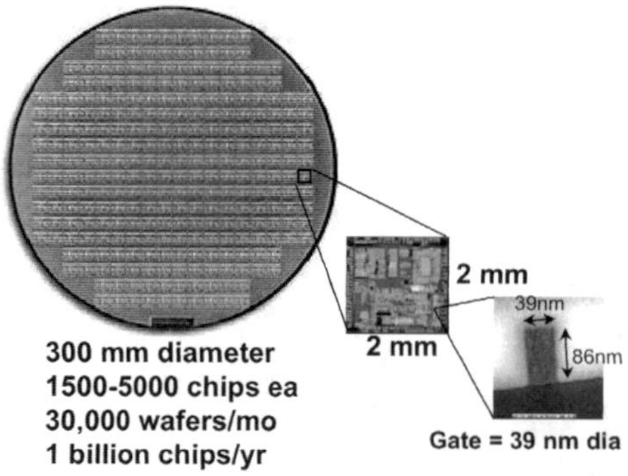

300 mm diameter
1500-5000 chips ea
30,000 wafers/mo
1 billion chips/yr

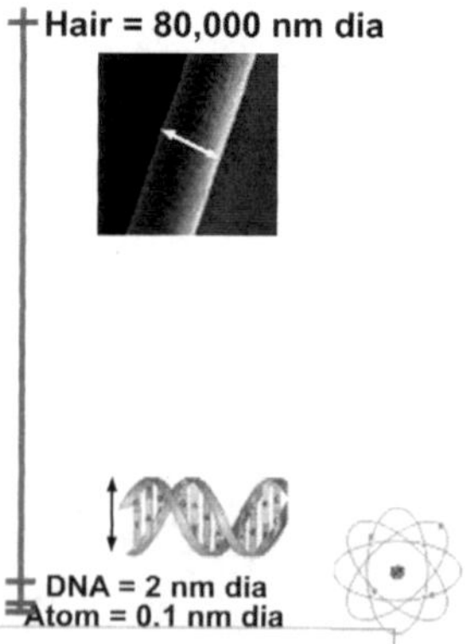

TEXAS INSTRUMENTS

Measurement

- Does anyone have a sustainable measuring device?
- Measurement is easier when broken into the major components:
  - Planet Factors
    - Energy
    - CO_2
    - Emissions
    - Water
    - Materials
    - Waste

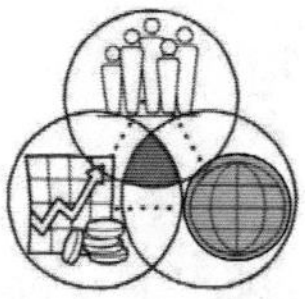

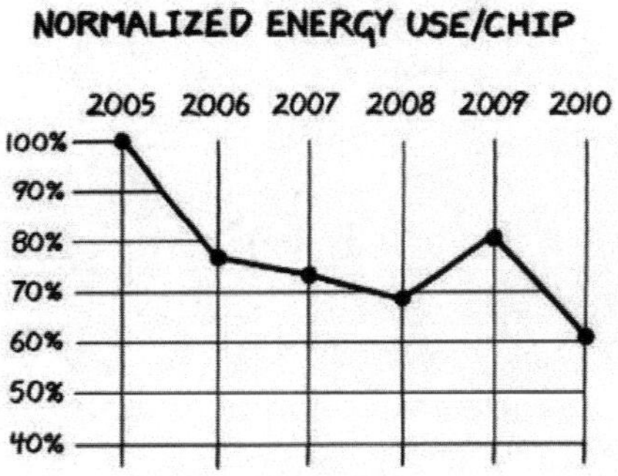

- Data can be normalized against square footage, units produced, weather, etc.
- Data is tracked over time and shared with employees to engage their assistance

17

Example Case: Semiconductor Fab

- 92-acre site
- 1.1 million square feet
- 220,000 square feet of cleanroom space
- Capacity for about 1,000 employees
- Main components:
  - Fab building (Fab=fabrication)
  - Support building
  - Central utility plant (CUP) building
  - Administration building

The Opportunity

- Very tight temperature and humidity requirements . . .
  - 70F+/-2 (21C+/-1) and 45% RH +/- 3%
- Combined with a large amount of exhaust and subsequent make up air . . .
  - 650,000 cfm (307 m3/sec) = 2 Macy's Kermit balloons per second

- Combined with the need to recirculate a large volume of air through the filters for cleanliness . . .
  - 4,400,000 cfm (2077 m3/sec) = 22 Goodyear blimps a minute

- Combined with hundreds of process tools with vacuum pumps, RF generators, and support equipment . . .
- Combined with extensive use of deionized (DI) water to rinse the wafers during processing . . .

Could lead to annual power consumption of 170,000 mWh (10,000 homes worth)
and water consumption of 3 million gallons/day (6,000 homes worth).
Annual utility bills could total $20M - $25M.

Texas Instruments

Strategy

- Strategy Team - Fabscape
  - 4 strategy teams were formed in advance of project
  - Request made to add a 5th team - sustainability
  - Generated early white papers on a number of ideas

- Tour My House (enerjazz.com/house)
  - Invited 3 TI VP's to tour active/passive solar home
  - Low utility bills for "normal" house spurred interest

- Design Workshop
  - Teamed up with Rocky Mountain Institute (RMI)
  - Held 3-day design charrette to brainstorm ideas
  - Generated 15 "Big Honkin' Ideas" to carry forward along with a large list of other good ideas
  - Made a first pass at LEED score sheet

Rocky Mountain Institute
abundance by design™

Texas Instruments

Project Management

- Sustainable Development Manager role assigned
  - Chief integrator of ideas and systems
  - The building was the system to be optimized
- Project Manager was fully supportive
  - Safety, Cost, Schedule, and Sustainability were front page
- Construction Contractor was the general contractor with the design firm working under them
  - Everyone on board from the start

Don't optimize the parts and pessimize the system

TEXAS INSTRUMENTS

Cost Reduction – Friend or Foe?

- Challenge:

  - Reduce project costs per square foot by 30 percent from the previous semiconductor fab
- Response:
  - A LEED building does not have to cost more
  - Forced space efficiency
  - Forced us to question everything
  - Couldn't just copy previous design – had to innovate
  - All of this led to . . . . Engineering!

TEXAS INSTRUMENTS

Integrated Design

You know you are on the right track when your solution for one problem accidentally solves several others.

- Amory Lovins, RMI

TEXAS INSTRUMENTS

Integrated Design Example

Office Lighting

Standard cost = $125
op cost = $40/yr

Ergolight cost = $375
op cost = $25/yr

Simple payback = 16.7 years 

However, we need 30% fewer Ergolight fixtures. Simple payback down to 6.7 years.

Efficient lighting also saves cooling energy. Simple payback down to 6.0 years.

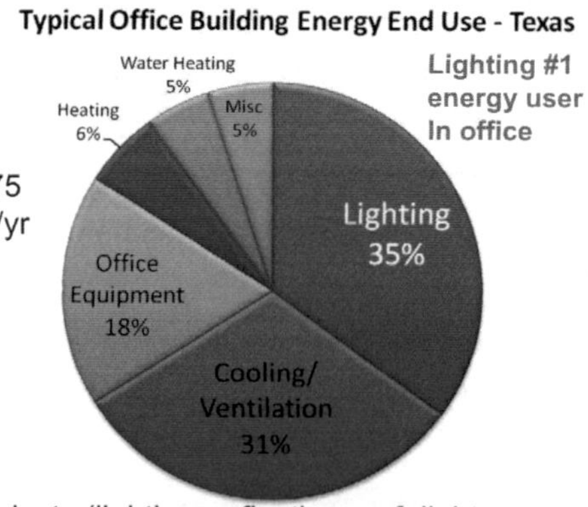

Add the contribution from dozens of similar projects (lighting, reflective roof, light shelves, sun shades, quality windows, extra insulation, better vacuum pumps, . .)

Enough cooling load disappears to avoid buying a $1M chiller . . . and the cooling tower, pumps, pipes, and even the space needed to install it.

Simple payback is now 0.0 years. The total net capital cost is the same, or even less, and the operating costs are lowered forever.

TEXAS INSTRUMENTS

Integrated Design Example

But wait, it gets even better . . .

Ergolights are individually controllable by each employee
Natural daylighting has been shown to increase productivity

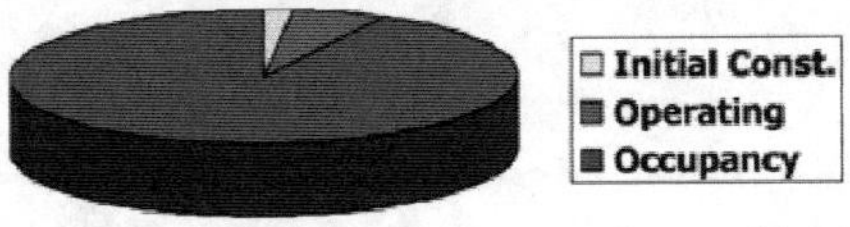

Cost of operation over 30 years for an office building

People costs account for 92% of all costs over a 30 year period.

If natural daylighting, self-control of lighting, improved indoor air quality, and all the other green building factors improved productivity by just 1% that would save the company >$1M/year for a large office complex.

Plus, if people like the building and control over their space it can give companies a recruiting advantage for top talent.

Texas Instruments

Office Efficiency

- Passive solar orientation with exterior shading
- Optimized glazing (high VLT, low SHGC, low U value)
- Reflective roof
- Natural daylighting with light shelves
- High-efficiency lighting (motion and daylight sensors)
- Demand-controlled ventilation
- Attention to detail on insulation and infiltration
- Solar water heating

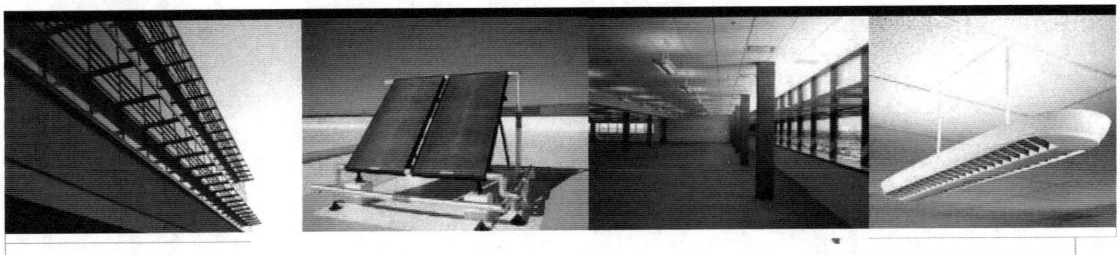

More Integration

< Areas were restored to native prairie grass to minimize irrigation and provide biodiversity.

Pond collects runoff from most of the 92 acres. 2.7 million gallon (10.2 million liters) base + 2 million gallon buffer. The pond meters runoff and settles sediment. Pond water is used for all site irrigation.

Windmill drives an air compressor to aerate the pond.

TEXAS INSTRUMENTS

Sustainability at RFAB

Native Meadow Restoration

Rain Water Reuse Pond

Reflective Roof

Dark Skies Friendly Lighting

Exterior Shades

Day lighting

Efficient Lights

Solar Water Heating

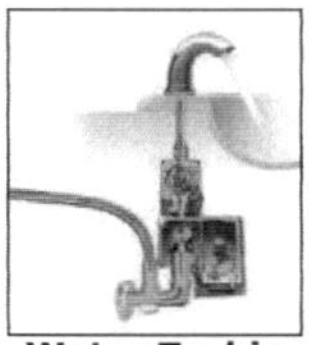

Water Turbine Powered Faucet

Bicycle Parking

Efficient cooling system with waste heat recovery

TEXAS INSTRUMENTS

RFAB Results

- We invested less than 1 percent of the project cost ($1.5M/$320M) in LEED-related items:
  - Predominately efficiency improvements that we would have considered regardless of LEED
- Overall project cost was 30 percent less per square foot than our previous fab just 6 miles away
- In the first year, we saved >$1 million in operating costs
- At full build out, we will save more than $4 million per year:
  - 25 percent energy reduction (have now reached 39%)
  - 40 percent water-use reduction
  - 50 percent on site emissions reduction
- LEED Gold Certified Office and Fab (first LEED Gold fab in the world)

www.ti.com/rfab

TEXAS INSTRUMENTS

TI Results

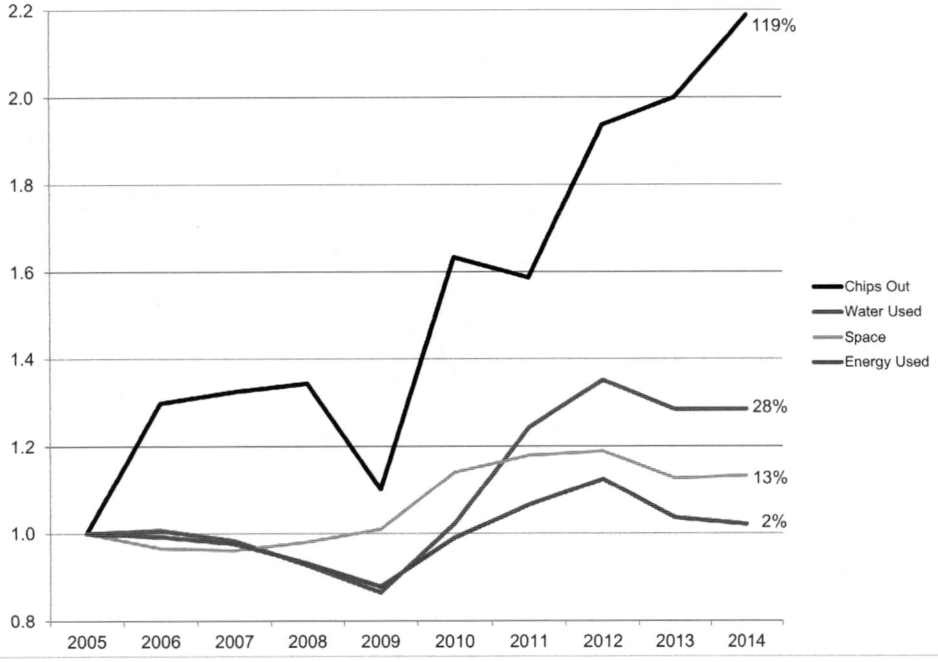

TEXAS INSTRUMENTS

Summary

- Develop the basic project requirements
- Set the sustainable goals early
- Organize for success
  - Have a broad team on board early
  - Assign a champion and integrator
  - Have a design session to exchange ideas, stretch thinking, and make connections
- Let form follow function
- Iterate and optimize the design
- Have multiple team members (example: the estimator and construction contractor) on board from the start
- Make sustainability a front page priority

TEXAS INSTRUMENTS

Bonus Ideas

- Design to operational performance goals – beyond code.
- Structure the contract to award / penalize for actual building energy and water efficiency performance
- Structure management bonuses contingent on reaching sustainable goals

TEXAS INSTRUMENTS

Don't Wait . . . Integrate

$\int_{Start}^{Finish}$ Design = Better Results

TEXAS INSTRUMENTS

Thank You

- Links
  - TI Citizenship Report – www.ti.com/ccr
  - TI RFAB – www.ti.com/rfab

  - Westbrook House – www.enerjazz.com/house
  - Paul's Prius – www.enerjazz.com/prius

Paul Westbrook
Sustainable Development Manager, Senior Member of the Technical Staff
Texas Instruments Facilities
Senior Fellow – US State Department Energy & Climate Partnership of the Americas (ECPA)
LEED AP

TEXAS INSTRUMENTS

15 AIChE
Spring Meeting, Austin, TX

Computational Modeling of Mixed Solids for CO_2 Capture Sorbents

Yuhua Duan
National Energy Technology Laboratory, United States Department of Energy
626 Cochrans Mill Road, Pittsburgh, Pennsylvania 15236, USA
Yuhua.duan@netl.doe.gov

Prepared for Presentation at
American Institute of Chemical Engineers
2015 Spring Meeting
11th Global Congress on Process Safety
Austin, Texas
April 27-30, 2015

UNPUBLISHED

Computational Modeling of Mixed Solids for CO_2 Capture Sorbents

Yuhua Duan
National Energy Technology Laboratory, United States Department of Energy
626 Cochrans Mill Road, Pittsburgh, Pennsylvania 15236, USA
Yuhua.duan@netl.doe.gov

Keywords: CO_2 solid sorbents, *ab initio* thermodynamics, first-principles density functional theory, lattice phonon dynamics, computational screening

Abstract

Since current technologies for capturing CO_2 to fight global climate change are still too energy intensive, there is a critical need for development of new materials that can capture CO_2 reversibly with acceptable energy costs. Accordingly, solid sorbents have been proposed to be used for CO_2 capture applications through a reversible chemical transformation. By combining thermodynamic database mining with first principles density functional theory and phonon lattice dynamics calculations, a theoretical screening methodology to identify the most promising CO_2 sorbent candidates from the vast array of possible solid materials has been proposed and validated. The calculated thermodynamic properties of different classes of solid materials versus temperature and pressure changes were further used to evaluate the equilibrium properties for the CO_2 adsorption/desorption cycles. According to the requirements imposed by the pre- and post- combustion technologies and based on our calculated thermodynamic properties for the CO_2 capture reactions by the solids of interest, we were able to screen only those solid materials for which lower capture energy costs are expected at the desired pressure and temperature conditions. Only those selected CO_2 sorbent candidates were further considered for experimental validations. The *ab initio* thermodynamic technique has the advantage of identifying thermodynamic properties of CO_2 capture reactions without any experimental input beyond crystallographic structural information of the solid phases involved. Such methodology not only can be used to search for good candidates from existing database of solid materials, but also can provide some guidelines for synthesis new materials. In this presentation, we apply our screening methodology to mixing solid systems to adjust the turnover temperature to help on developing CO_2 capture Technologies.

1. Introduction

Nowadays, the burning of fossil fuels is still the main energy source for the world's economy. One consequence of the use of these fuels is the emission of huge quantities of CO_2 into the atmosphere creating environmental problems such as global climate warming.[1-5] In order to solve such problems, there is a need to reduce CO_2 emissions into atmosphere by capturing and sequestrating CO_2.[4, 6, 7] One approach to solve such environmental problems is to capture and sequester the CO_2. Current technologies for capturing CO_2 including solvent-based (amines) and CaO-based materials are still too energy intensive. Hence, there is critical need for development of new materials that can capture and release CO_2 reversibly with acceptable energy costs. In particular, solid oxide sorbent materials have been proposed for capturing CO_2

Note: Do not add page numbers. Do not refer to page numbers when referencing different portions of the paper

through a reversible chemical transformation leading primarily to formation of carbonate products. Solid sorbents containing alkali and alkaline earth metals have been reported in several previous studies to be promising candidates for CO_2 sorbent applications due to their high CO_2 absorption capacity at moderate working temperatures.[8-11]

During past few years we developed a theoretical methodology to identify promising solid sorbent candidates for CO_2 capture by combining thermodynamic database searching with *ab initio* thermodynamics obtained based on first-principles density functional theory (DFT) and lattice phonon dynamics.[8-10, 12-16] As shown in Fig.1, the primary outcome of our screening scheme is a list of promising CO_2 sorbents with optimal energy usage.

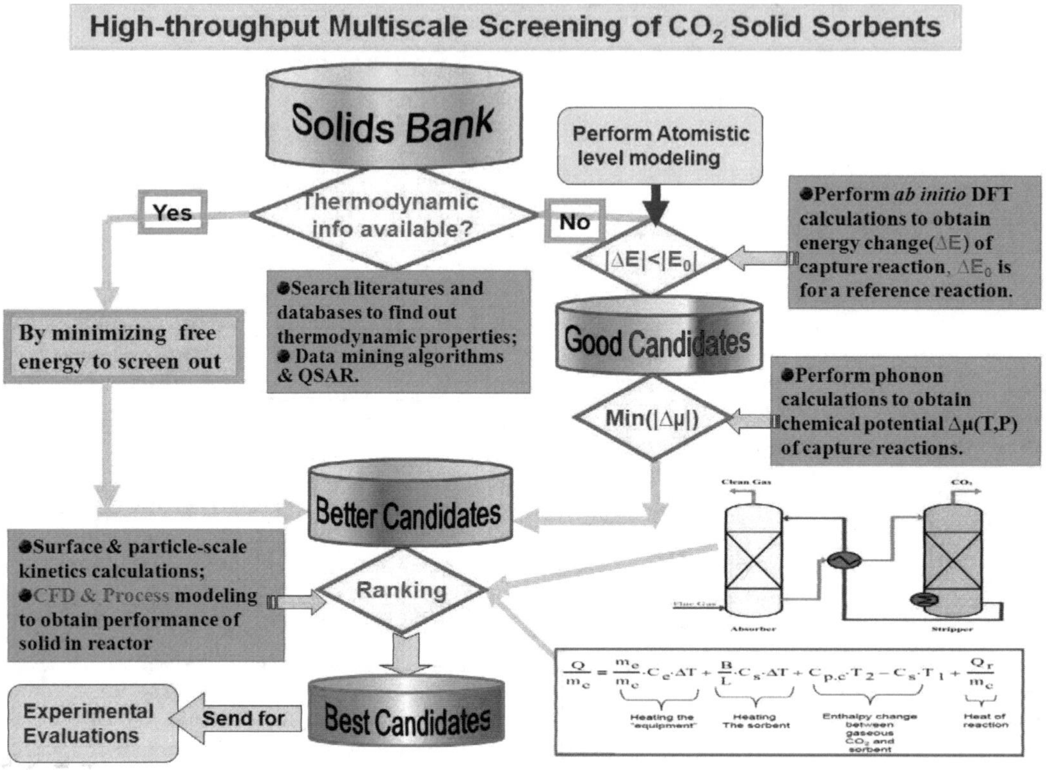

Fig. 1. Schematic of our screening methodology.

At a given CO_2 pressure, the turnover temperature (T_t) of an individual solid capture CO_2 reaction is fixed. Such T_t may be outside the operating temperature range (ΔT_o) for a particularly capture technology. In order to adjust T_t to fit the practical ΔT_o, in this study, we demonstrate that by mixing different types of solids it's possible to shift T_t to the range of practical operating conditions.

2. Calculation Methods for Mixed Solid Sorbents

The complete description of the computational methodology together with relevant applications can be found in our previous publications.[8-10, 12-19] The CO_2 capture reactions of solids can be expressed generically in the form (for convenient description, we normalized the reaction to 1 mole of CO_2)

$$\sum_{Ri} n_{Ri}\text{Solid}\_R_i + CO_2 \leftrightarrow \sum_{Pj} n_{Pj}\text{Solid}\_P_j \qquad (a)$$

where n_{Ri}, n_{Pj} are the numbers of moles of reactants (R_i) and products (P_j) involved in the capture reactions. We treat the gas phase CO_2 as an ideal gas. By assuming that the difference between the chemical potentials ($\Delta\mu^o$) of the solid phases of reactants (R_i) and products (P_j) can be approximated by the difference in their total energies (ΔE^{DFT}), obtained directly from DFT calculations, and the vibrational free energy of the phonons and by ignoring the PV contribution terms for solids, the variation of the chemical potential ($\Delta\mu$) for reaction (a) with temperature and pressure can be written as[8-10, 12-19]

$$\Delta\mu(T,P) = \Delta\mu^0(T) - RT\ln\frac{P_{CO_2}}{P_0} \qquad (1)$$

where

$$\Delta\mu^0(T) \approx \Delta E^{DFT} + \Delta E_{ZP} + \Delta F^{PH}(T) - G^0_{CO_2}(T) \qquad (2)$$

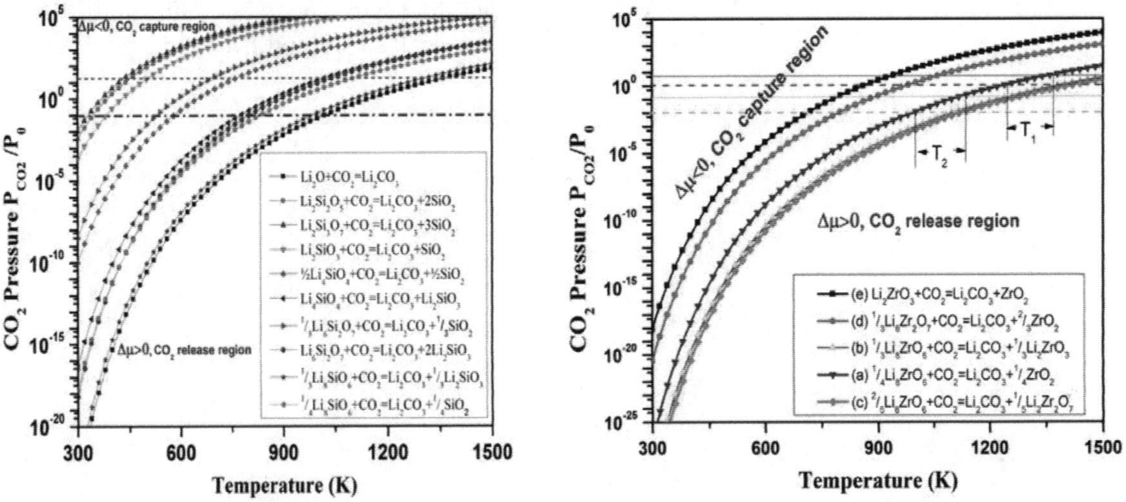

(a) Li_2O/SiO_2 (b) Li_2O/ZrO_2

Fig. 2. Contour plots of the calculated chemical potential ($\Delta\mu$) versus temperature and the CO_2 pressure (*P* plotted in logarithmic scale) for the CO_2 capture reactions. Only $\Delta\mu=0$ curve is shown explicitly. For each reaction, above its $\Delta\mu=0$ curve, their $\Delta\mu<0$, which means the sorbents absorb CO_2 and the reaction goes forward, whereas below the $\Delta\mu=0$ curve, their $\Delta\mu>0$, which indicates CO_2 start to be released and reaction goes backward with regeneration of the sorbents.

Here, ΔE^{DFT} is the DFT energy difference between the reactants and products of the reaction (a), ΔE_{ZP} is the zero point energy difference between the reactants and products and can be obtained directly from phonon calculations. ΔF^{PH} is the phonon free energy change excluding zero-point energy (which is already counted into the ΔE_{ZP} term) between the solids of products and reactants. P_{CO2} is the partial pressure of CO_2 in the gas phase and P_0 is the standard state reference pressure taken to be 1 bar. The heat of reaction ($\Delta H^{cal}(T)$) can be evaluated through the following equation

$$\Delta H^{cal}(T) = \Delta\mu^0(T) + T[\Delta S_{PH}(T) - S_{CO_2}(T)] \qquad (3)$$

where $\Delta S_{PH}(T)$ is the difference of entropies between product solids and reactant solids. The free energy of CO_2 (G^0_{CO2}) can be obtained from standard statistical mechanics,[9, 10, 16] and its entropy (S_{CO2}) can be found in the empirical thermodynamic databases.[20]

3. Results and Discussions

For a given CO_2 capture process, its optimal working conditions (CO_2 pressures of pre- and after-capture, absorption/desorption temperature range (ΔT_o), *etc.*) were fixed. However, at a given CO_2 pressure, the turnover temperature (T_t) of an individual solid capture CO_2 reaction is fixed. Such T_t may be outside the operating temperature range ΔT_o for a particularly capture technology. In order to adjust T_t to fit the practical working through reversible chemical transformations ΔT_o, we have demonstrated that by mixing different types of solids it's possible to shift T_t to the practical operating ΔT_o range.[8, 21, 22] Generally, when we mix two solids *A* and *B* to form a new sorbent *C*, the turnover temperature of the newly resulted system (T_C) is located between those of *A* and *B* (T_A, T_B). Here it was assumed that *A* is a strong CO_2 sorbent while *B* is a weak CO_2 sorbent and $T_A>T_B$. Also, we assumed that the desired operating temperature T_O is between T_A and T_B ($T_A>T_O>T_B$). Depending on the properties of *A* and *B*, we have typically three scenarios to synthesize the mixing sorbent *C*:

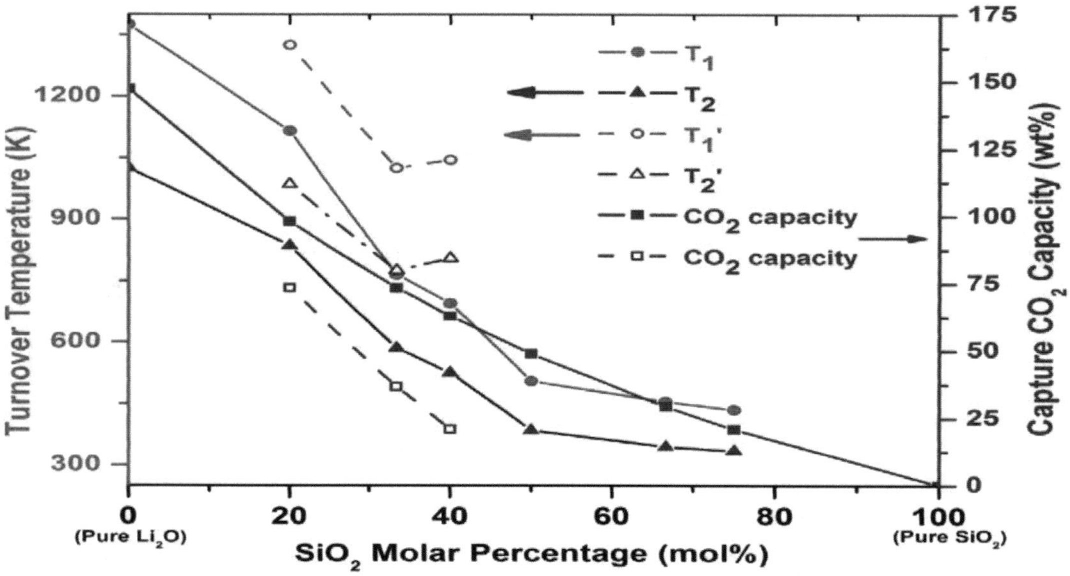

Fig. 3 The dependence of the turnover temperatures defined in the text and of CO_2 capture capacity on molar percentage of SiO_2 in the silicates for which calculations are reported here. T_1 and T_1' are the turnover temperatures under pre-combustion conditions with CO_2 partial pressure at 20 bars, while T_2 and T_2' are the turnover temperatures under post-combustion conditions with CO_2 partial pressure at 0.1 bar. The solid lines indicate to convert lithium silicates to SiO_2 and Li_2CO_3 (T_1, T_2). For those Li_2O-rich lithium silicates (Li_8SiO_6, Li_4SiO_4, $Li_6Si_2O_7$) capturing CO_2, the data shown in dash lines indicate convert them to Li_2SiO_3 and Li_2CO_3 (T_1', T_2'). The corresponding CO_2 capture capacities are plotted with open blue squares.

3.1. $T_A >> T_B$ and the *A* component is the key part to capture CO_2.

An example of this case is represented by Li_2O. This is a very strong CO_2 sorbent which forms Li_2CO_3. However, its regeneration from Li_2CO_3 only can occur at very high temperature (T_A). In order to move its T_A to lower temperatures, one can mix some weak CO_2 sorbents (such as SiO_2, ZrO_2). Fig. 2 shows the relationship of chemical potential, P_{CO2}, and T of the CO_2 capture reactions by the mixed Li_2O/SiO_2 and Li_2O/ZrO_2 solids with different mixing ratios. Fig. 3 shows the turnover temperature and the CO_2 capture capacity of Li_2O/SiO_2 mixture versus the ratio of Li_2O/SiO_2.[8-10, 12, 18, 21, 23, 24]

From Figs. 2 and 3, one can see that after mixing Li_2O and SiO_2 (or ZrO_2) with different Li_2O/SiO_2 (or Li_2O/ZrO_2) ratios, the T_C of the newly formed *C* compound (silicate or zirconate) is lower than T_A of pure Li_2O and could be close to the ΔT_o range to fit the practical needs.

3.2 $T_A >> T_B$ and B component is the key part to capture CO_2

In this case, since T_B is lower than T_O, by mixing *A* into *B* will increase the turnover temperature T_C of the *C* solid to values closer to T_o. For example, pure MgO has a very high theoretical CO_2 capture capacity. However, its turnover temperature (250 °C) is lower than the required temperature range of 300-470 °C used in warm gas clean up technology and its practical CO_2 capacity is very low, and therefore, it cannot be used directly as a CO_2 sorbent in this technology.[25-27] As shown in Fig. 4, by mixing alkali metal oxides M_2O (M=Na, K, Cs, Ca) or carbonates (M_2CO_3) into MgO, the corresponding newly formed mixing systems have higher turnover temperatures making them useful as CO_2 sorbents through the reaction $MgO + CO_2 + M_2CO_3 = M_2Mg(CO_3)_2$.[25, 28]

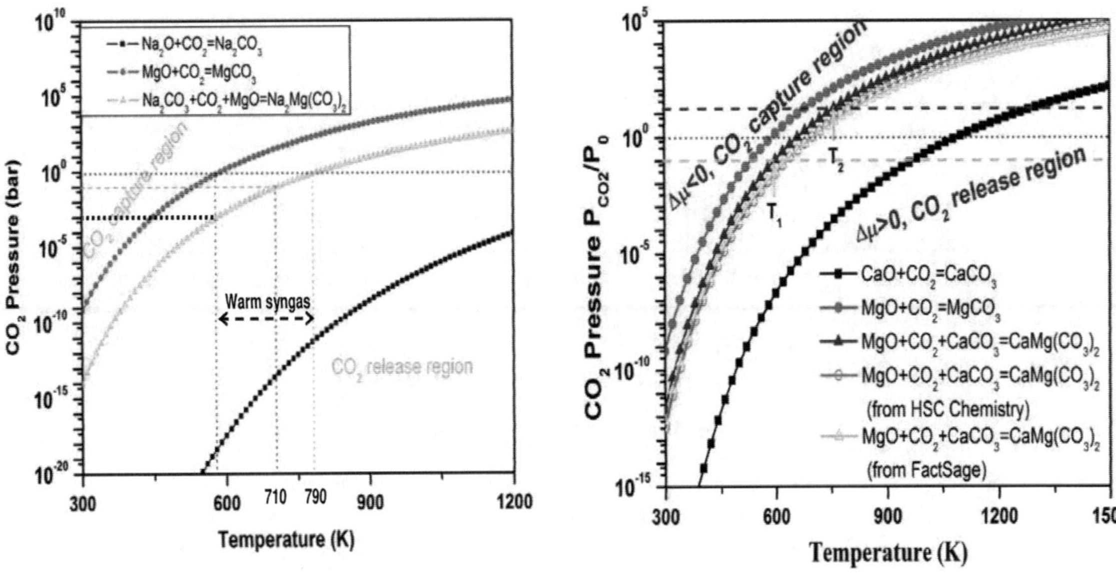

(a) MgO/Na_2CO_3 (Na_2O) (b) $MgO/CaCO_3$ (CaO)

Fig. 4. Plots of the calculated chemical potentials versus CO_2 pressures and temperatures for the CO_2 capture reaction by MgO to form double salts.

3.3 T_A and T_B are close to each other and both A and B are active for CO_2 capture

In this case, both *A* and *B* components are active to capture CO_2, and the CO_2 capacity of the mixture is the summation of those of *A* and *B*. As we know another potential advantage of mixing solids is to increase the surface area of the solids in order to have faster reaction rates. Such a mixing scenario doesn't show too much advantage in shifting the capture temperature, but may enhance the kinetics of the capture process and eventually make the mixtures more efficient. Although there is no such report in literature, we think such an attempt is worthwhile and are working on several doped systems. Figure 5 shows a case of Na-/K- doped Li_2ZrO_3 capture CO_2 through the following reaction (b):[29]

$$LiMZrO_3 + CO_2 \leftrightarrow \frac{1}{2}Li_2CO_3 + \frac{1}{2}M_2CO_3 + ZrO_2 \qquad (M = Na, K) \qquad (b)$$

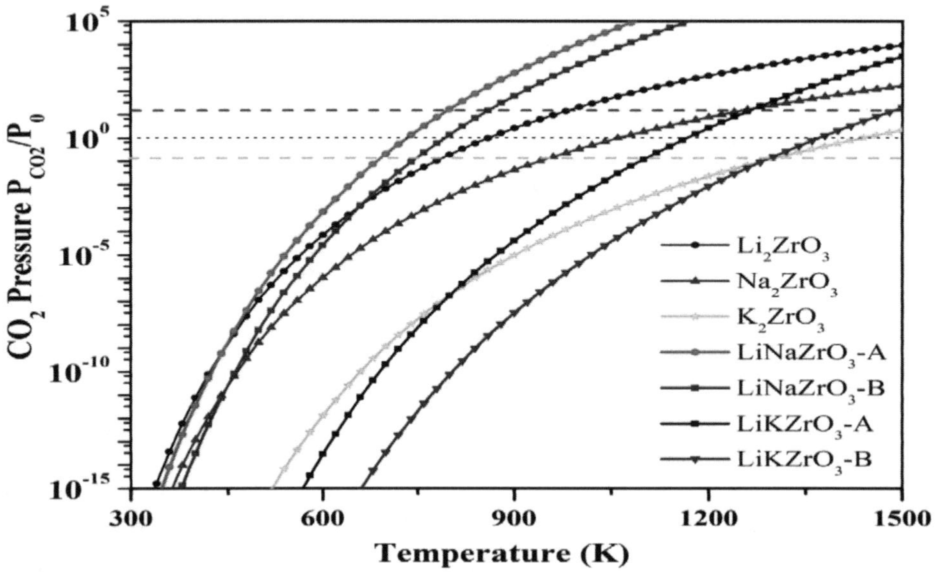

Fig. 5. The contour plotting of calculated chemical potentials versus CO_2 pressures and temperatures of the CO_2 capture reactions by $LiMZrO_3$ and M_2ZrO_3. Y-axis plotted in logarithm scale. Only the $\Delta\mu=0$ curve is shown explicitly. For each reaction, above its $\Delta\mu=0$ curve, their $\Delta\mu<0$, which means the solids absorb CO_2 and the reaction goes forward, whereas below the $\Delta\mu=0$ curve, their $\Delta\mu>0$, which means the CO_2 start to release and the reaction goes backward to regenerate the sorbents.

All of the reactions are thermodynamically favorable for CO_2 sorption over a quite wide range of temperatures and reduced CO_2 pressures. But a well preforming CO_2 solid sorbent should not only easily absorb CO_2 in the first half cycle but also easily release CO_2 from the product (M_2CO_3 and ZrO_2 for example) in the second half cycle. Table 1 shows the calculated thermodynamical properties and turnover temperatures of $LiMZrO_3$ capture CO_2 reaction (b).

From Fig.5 and Table 1, one can see that they are not good sorbents for capturing CO_2 for pre-combustion technology. However, they could be used for high-temperature post-combustion CO_2 capture with T_2=685K, 745 K for $LiNaZrO_3$ (phases _A and _B) and T_2=1095K, 1285 K for $LiKZrO_3$ (phases _A and _B) respectively. When compared to Li_2ZrO_3, the K substituted system

$LiKZrO_3$ does not decrease the turnover temperature of the CO_2 capture reaction. In other words, although the $LiKZrO_3$ has a stronger driving force to capture CO_2, with improved CO_2 capture capability when compared to Li_2ZrO_3, its regeneration only can happen at higher temperature. Therefore, $LiKZrO_3$ is not a good CO_2 sorbent, since it requires more energy to regenerate. In contrast, the Na-substituted sorbent system $LiNaZrO_3$ does have a more compatible turn over temperature, making it a possible good candidate as a CO_2 sorbent. Such conclusions are in good agreement with experimental findings.[30-32]

In summary, from the calculated thermodynamic properties of $LiMZrO_3$ reacting with CO_2, we found that the CO_2 sorbent performance of $LiNaZrO_3$ is better than that of Li_2ZrO_3, but the performance of $LiKZrO_3$ is much worse since its regeneration temperature is higher. Therefore, the Na doped Li_2ZrO_3, $LiNaZrO_3$, is a better CO_2 sorbent for post-combustion capture technology. Our calculated thermodynamic results showed that the K doped Li_2ZrO_3, $LiKZrO_3$, didn't gain any improvements over Li_2ZrO_3 on its overall CO_2 capture performance.

Table 1. The weight percentage of CO_2 capture, the calculated energy change ΔE^{DFT}, the zero-point energy changes ΔE^{ZP} and the thermodynamic properties (ΔH, ΔG) of the CO_2 capture reactions. (unit: kJ/mol). The turnover temperatures (T_1 and T_2) of the reactions of CO_2 capture by solids under the conditions of pre-combustion (P_{CO2}=20 bar) and post-combustion (P_{CO2}=0.1 bar) are also listed.

reaction	absorbing CO_2 Wt%	ΔE^{DFT}	ΔE^{ZP}	ΔH (T=300K)	ΔG (T=300K)	Turnover T (K)	
						T_1	T_2
$Li_2ZrO_3+CO_2\leftrightarrow Li_2CO_3+ZrO_2$ [a]	28.75	-146.648	11.311	-158.562	-103.845	1000	780
$LiNaZrO_3$_A$+CO_2\leftrightarrow$½$(Li_2CO_3+Na_2CO_3)+ZrO_2$	26.01	-152.936	7.069	-176.666	-110.892	805	685
$LiNaZrO_3$_B$+CO_2\leftrightarrow$½$(Li_2CO_3+Na_2CO_3)+ZrO_2$	26.01	-167.872	6.934	-191.526	-126.477	865	745
$LiKZrO_3$_A$+CO_2\leftrightarrow$½$(Li_2CO_3+K_2CO_3)+ZrO_2$	23.75	-264.115	6.006	-287.513	-225.611	1275	1095
$LiKZrO_3$_B$+CO_2\leftrightarrow$½$(Li_2CO_3+K_2CO_3)+ZrO_2$	23.75	-311.604	7.080	-332.612	-272.410	hT[b]	1285
$Na_2ZrO_3+CO_2\leftrightarrow Na_2CO_3+ZrO_2$ [a]	23.76	-140.862	2.236	-158.327	-114.121	1275	925
$K_2ZrO_3+CO_2\leftrightarrow K_2CO_3+ZrO_2$ [a]	20.24	-223.158	5.813	-238.490	-187.884	hT[b]	1285

[a] from Refs.[18, 23, 24].
[b] hT means the temperature is higher than our temperature range (1500K)

With similar approach, we have investigated the electronic structural and phonon properties of $Na_{2-\alpha}M_\alpha ZrO_3$ (M=Li, K, α=0, 0.5, 1.0, 1.5, 2.0) by density functional theory and first-principles determination of phonon dynamics.[33] The thermodynamic properties of CO_2 absorption/desorption in these materials were analyzed. The obtained results are presented in Fig.6 and Table 2.

Based on the calculated relationships among the chemical potential change, CO_2 pressure and temperature for CO_2 capture reactions by $Na_{2-\alpha}M_\alpha ZrO_3$ and thermogravimetric analysis experimental measurements, compared to pure Na_2ZrO_3, the Li- and K-doped mixtures $Na_{2-\alpha}M_\alpha ZrO_3$ have lower turnover temperatures (T_t) and higher CO_2 capacities. The calculated results show that the shift in T_t depends not only on the doping element, but also depends on the doping level. The Li-doped systems have larger T_t decreases than the K-doped systems. When increasing the Li-doping level α, the T_t of corresponding mixture $Na_{2-\alpha}Li_\alpha ZrO_3$ decreases further to a low temperature range. However in the case of K-doped systems $Na_{2-\alpha}K_{\alpha}ZrO_3$, although initial doping of K into Na_2ZrO_3 can shift its T_t to lower temperature range, further increasing the K-doping level α results in an increase in T_t. Therefore, compared to K-doing, lithium inclusion into Na_2ZrO_3 structure has a larger influence on the CO_2 capture performance.

Table 2. The weight percentage of CO_2 capture, the ratios of Na_2O:M_2O:ZrO_2 (M=Li, K), the calculated energy change ΔE^{DFT}, the zero-point energy changes ΔE^{ZP} and the thermodynamic properties (ΔH, ΔG) of the CO_2 capture reactions. (unit: kJ/mol). The turnover temperatures (T_1 and T_2) of the reactions of CO_2 capture by solids under the conditions of pre-combustion (P_{CO2}=20 bar) and post-combustion (P_{CO2}=0.1 bar) are also listed.[33]

reaction	absorbing CO_2 wt%	Na_2O:M_2O:ZrO_2 ratio	ΔE^{DFT}	ΔE^{ZP}	ΔH (T=300K)	ΔG (T=300K)	Turnover T (K)	
							T_1	T_2
$Na_2ZrO_3+CO_2 \leftrightarrow Na_2CO_3+ZrO_2$ [a]	23.76	1:0:1	-140.862	2.236	-158.327	-114.121	1275	925
$Na_{1.5}Li_{0.5}ZrO_3$-B+$CO_2 \leftrightarrow {}^3/_4Na_2CO_3+{}^1/_4Li_2CO_3+ZrO_2$	24.83	³/₄:¹/₄:1	-170.881	4.667	-242.090	-159.144	805	715
$Na_{1.0}Li_{1.0}ZrO_3$-B+$CO_2 \leftrightarrow {}^1/_2Na_2CO_3+{}^1/_2Li_2CO_3+ZrO_2$	26.01	½:¹/₂:1	-157.839	6.480	-228.381	-142.555	745	675
$Na_{0.5}Li_{1.5}ZrO_3$-A+$CO_2 \leftrightarrow {}^1/_4Na_2CO_3+{}^3/_4Li_2CO_3+ZrO_2$	27.31	¼:³/₄:1	-169.827	9.652	-237.765	-146.230	735	665
$Na_{1.5}K_{0.5}ZrO_3$-A+$CO_2 \leftrightarrow {}^3/_4Na_2CO_3+{}^1/_4K_2CO_3+ZrO_2$	22.77	³/₄:¹/₄:1	-210.081	2.486	-281.253	-199.996	915	825
$Na_{1.0}K_{1.0}ZrO_3$-B+$CO_2 \leftrightarrow {}^1/_2Na_2CO_3+{}^1/_2K_2CO_3+ZrO_2$	21.86	½:¹/₂:1	-245.436	2.058	-316.736	-236.789	1015	915
$Na_{0.5}K_{1.5}ZrO_3$-B+$CO_2 \leftrightarrow {}^1/_4Na_2CO_3+{}^3/_4K_2CO_3+ZrO_2$	21.02	¼:³/₄:1	-278.147	1.519	-349.077	-272.038	1125	1015
$Li_2ZrO_3+CO_2 \leftrightarrow Li_2CO_3+ZrO_2$ [b]	28.75	0:1:1	-146.648	11.311	-158.562	-103.845	1000	780
$K_2ZrO_3+CO_2 \leftrightarrow K_2CO_3+ZrO_2$ [a]	20.24	0:1:1	-223.158	5.813	-238.490	-187.884	hT[c]	1285

[a] From Ref.[24].
[b] From Ref.[18]
[c] hT means the temperature is higher than 1500K

All these results may be of great interest in the development of specific CO_2 capture applications. As it has been shown, the $Na_{2-\alpha}Li_{\alpha}ZrO_3$ and $Na_{2-\alpha}K_{\alpha}ZrO_3$ compositions can produce modifications in the CO_2 capture temperatures, which may be used in the design of a specific composition depending on the temperature range that industry requires. Our work has identified that the capture of CO_2 in zirconate materials is not simply a matter of substitutional element, but also the doping level. This insight will need to be considered during future sorbent development. We have also demonstrated that computational methods can be used to accurately predict aspects of CO_2 capture and have the potential to drive future work by identifying the most promising candidate materials.

4. Conclusions

The obtained results showed that by changing the mixing ratio of solid *A* and solid *B* to form mixed solid *C* it's possible to shift the turnover T_t of the newly formed solid *C* to fit the practical CO_2 capture technologies. When mixing SiO_2 or ZrO_2 into the strong Li_2O sorbent, one can obtain a series of lithium silicates (or zirconates) with T_t lower than that of pure Li_2O. By mixing oxides (Na_2O, K_2O, CaO) or their corresponding carbonates into MgO, the obtained mixtures exhibit different thermodynamic behaviors and their T_t are higher than that of pure MgO. Such results can be used to provide insights for designing new CO_2 sorbents. Therefore, although one single material taken in isolation might not be an optimal CO_2 sorbent to fit the particular needs to operate at specific temperature and pressure conditions, by mixing or doping two or more materials to form a new material, our results showed that it is possible to synthesize new CO_2 sorbent formulations which can fit the industrial needs. Our results also show that computational modeling can play a decisive role for identifying materials with optimal performance.

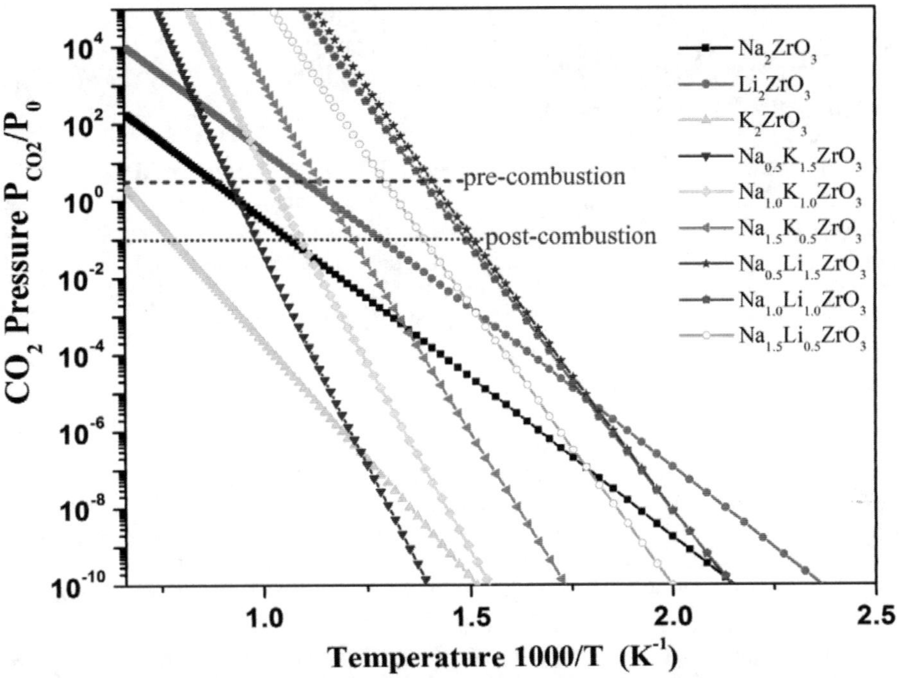

Fig. 6. The contour plotting of calculated chemical potential changes versus CO_2 pressures and temperatures of the reactions for $Na_{2-\alpha}M_{\alpha}ZrO_3$ (M=Li, K, α=0.5, 1.0, 1.5) capture CO_2. The typical CO_2 pressure for pre- and post-combustions are specified in blue lines.[33]

Acknowledgement

The author thanks Drs. D. C. Sorescu, D. Luebke, H. W. Pennline, B. Y. Li, H. Pfeiffer, J. K. Johnson, B. Zhang, K. Zhang, D. King, K. Parlinski for their kind help and collaborations.

References

[1] D. Aaron, and C. Tsouris, Separation Science and Technology **40**, 321 (2005).
[2] C. M. White, B. R. Strazisar, E. J. Granite, J. S. Hoffman, and H. W. Pennline, J Air Waste Manag Assoc **53**, 645 (2003).
[3] R. S. Haszeldine, Science **325**, 1647 (2009).
[4] B. Y. Li, Y. Duan, D. Luebke, and B. Morreale, Applied Energy **102**, 1439 (2013).
[5] M. R. Allen, D. J. Frame, C. Huntingford, C. D. Jones, J. A. Lowe, M. Meinshausen, and N. Meinshausen, Nature **458**, 1163 (2009).
[6] H. Pfeiffer, and P. Bosch, Chemistry of Materials **17**, 1704 (2005).
[7] E. Ochoa-Fernandez, H. K. Rusten, H. A. Jakobsen, M. Ronning, A. Holmen, and D. Chen, Catalysis Today **106**, 41 (2005).
[8] Y. Duan, D. Luebke, and H. W. Pennline, International Journal of Clean Coal and Energy **1**, 1 (2012).
[9] Y. Duan, and D. C. Sorescu, Phys Rev B **79**, 014301 (2009).
[10] Y. Duan, and D. C. Sorescu, J Chem Phys **133**, 074508 (2010).
[11] Y. Duan, in *Proceedings of 7th~13th Ann. Conf. on Carbon Capture, Sequestration & Utilization*Pittsburgh, 2008-2014).
[12] Y. Duan, and K. Parlinski, Phys Rev B **84**, 104113 (2011).
[13] Y. Duan, B. Zhang, D. C. Sorescu, and J. K. Johnson, J. Solid State Chem. **184**, 304 (2011).
[14] Y. Duan, D. R. Luebke, H. W. Pennline, B. Y. Li, M. J. Janik, and J. W. Halley, Journal of Physical Chemistry C **116**, 14461 (2012).
[15] Y. Duan, B. Zhang, D. C. Sorescu, J. K. Johnson, E. H. Majzoub, and D. R. Luebke, Journal of Physics-Condensed Matter **24**, 325501 (2012).
[16] B. Zhang, Y. Duan, and J. K. Johnson, J Chem Phys **136**, 064516 (2012).
[17] Y. Duan, Phys. Rev. B **77**, 045332 (2008).
[18] Y. Duan, J Renew Sustain Ener **3**, 013102 (2011).
[19] Y. Duan, J Renew Sustain Ener **4**, 013109 (2012).
[20] M. W. J. Chase, J. Phys. Chem. Ref. Data, **Monograph 9**, 1 (1998).
[21] Y. Duan, Physical Chemistry Chemical Physics **15**, 9752 (2013).
[22] Y. Duan, H. Pfeiffer, B. Li, I. C. Romero-Ibarra, D. C. Sorescu, D. R. Luebke, and J. W. Halley, Physical Chemistry Chemical Physics **15**, 13538 (2013).
[23] Y. Duan, H. Pfeiffer, B. Y. Li, I. C. Romero-Ibarra, D. C. Sorescu, D. R. Luebke, and J. W. Halley, Physical Chemistry Chemical Physics **15**, 13538 (2013).
[24] Y. Duan, J Renew Sustain Ener **4**, 013109 (2012).
[25] K. L. Zhang, X. H. S. Li, Y. Duan, D. L. King, P. Singh, and L. Y. Li, International Journal of Greenhouse Gas Control **12**, 351 (2013).
[26] K. L. Zhang, X. H. S. Li, W. Z. Li, A. Rohatgi, Y. Duan, P. Singh, L. Y. Li, and D. L. King, Advanced Materials Interfaces **1**, 1400030 (2014).
[27] J. L. Chi, L. F. Zhao, B. Wang, Z. Li, Y. H. Xiao, and Y. Duan, International Journal of Hydrogen Energy **39**, 6479 (2014).
[28] Y. Duan, K. Zhang, X. S. Li, D. L. King, B. Li, L. Zhao, and Y. Xiao, Aerosol and Air Quality Research **14**, 470 (2014).
[29] Y. Duan, ScienceJet **3**, 56 (2014).
[30] M. Y. Veliz-Enriquez, G. Gonzalez, and H. Pfeiffer, J Solid State Chem **180**, 2485 (2007).
[31] H. Pfeiffer, E. Lima, and P. Bosch, Chemistry of Materials **18**, 2642 (2006).
[32] H. Pfeiffer, C. Vazquez, V. H. Lara, and P. Bosch, Chemistry of Materials **19**, 922 (2007).
[33] Y. Duan, J. Lekse, X. F. Wang, B. Y. Li, B. Alcantar-Vazquez, H. Pfeiffer, and J. W. Halley, Physical Review Applied, accepted (2015).

399080 Zero Ammonia Gas Emmission Strategy @ Ammonia Urea Complex

Abstract:

Engro Fertilizers limited is 2nd largest Urea Producer in Pakistan. Engro is operating **02** Ammonia, **03** Urea and **03** Utilities plants near a rich natural gas facility in district Ghokti with total capacity of **2.1** Million tons per year. In 2014 Engro achieved land mark **DuPont 4.3** ranking in process safety management. High standard of process safety has been achieved through top management commitment and developing a marvelous system of integrated safety management, now Engro is striving for world class safety and Environment performer. Engro commissioned new Ammonia Urea complex in **2010** based on Snamprogetti technology having **2194 MeT/day** Liquid Ammonia and **3835 MeT/day** Urea capacity.

Engro Fertilizers HSE policy is true depiction of its top management commitment towards Process Safety and Environment Management. One clause of policy says *"Continuously improve environmental performance to achieve sustainable development."* In this regard state of the art technology, flare system was procured for new Ammonia Urea complex to ensure zero ammonia gas effluent during normal operation, start up, shutdown and process upset. This initiative eradicated the all possible risks related to ammonia release in environment. Flare system protected workforce at plant, surrounding community and trespassers. It has been designed to burn continuous ammonia effluent as well as to cater abrupt ammonia release in case of process upset. Since commissioning of new Ammonia Urea plant, environment complains have been drastically reduced and one element of environment pollution have been eradicated. Moreover detail PHA studies were conducted to analyze the relevant hazards and their consequences. After commissioning, a key in house design modification was conducted to avoid localized Ammonical water release at Urea plant, which not only avoided ammonia release from Urea plant permissive but also resulted in ammonia recovery through close header, recovered ammonia recycled to Urea process which enhanced Urea yield.

399820 Importance of Process and Flaring Data and Its Analysis for the Management of Abnormal Situations: An Ethylene Process Case Study

Abstract:

Upset emissions occur during plant startup, shutdown, maintenance, malfunction and flaring incidents. A wide range of these Upsets cannot be managed by the standalone control systems even the advanced ones and need the interfere of plant personnel to correct the causes of this abnormal situation. The technologies needed to help plant personnel to control and prevent abnormal situations are gathered under abnormal situation management. These situations have significant impact on the process industry economy, safety and environment. Moreover, flaring, which is a key but negative method for managing abnormal situations in petrochemical plants, emits huge amounts of greenhouse gases (GHG) and hazardous volatile organic compounds (VOC_S), causing locally transient air pollution and negative community impacts. In addition, flaring is causing huge loss in raw material and energy which could be used to generate more desired products for the industry. Thus, flare minimization during normal and abnormal operational situations has great benefits to environmental, societal, and industrial sustainability. It is not possible to quantify the impacts without understanding the properties and magnitude of these upsets. The analysis of the properties and magnitude of the upset emissions needs extensive amount of historical data. In this paper, an ethylene base case study along with its historical process and flaring incidents data are used to identify underlying causes and indicators that lead to process upsets, i.e. abnormal situations. The paper will also show potential technique(s) that are recommended to mitigate the incidents and reduce/eliminate its root causes. Moreover, a systematic approach to collect and analyze historical flaring data based on current industrial practices is presented. Finally, the challenges that are faced by environmental agencies in terms of data utilization and documentation are highlighted with some recommended improvements and suggestions.

396253 Studies on Briquetting of Chromite Ore with Waste Plastic Binder

Abstract:

This work focuses on the development of new binders with the help of waste plastics for briquetting of ore fines and to measure its cold compressive strength (CCS). Chromite is an important mineral, during the mining operation lot of fines are produced which cannot be easily used by existing technology because most of the technology are cost intensive. Briquetting is considered a cheap and easy method but it needs suitable binder. Plastic is usually synthetic and non-biodegradable material. Now a days a large quantity of plastic is being generated as a waste, utilization of these wastes are very difficult to carry out. One of the best use of these plastic waste is to develop as a binders for ore fines briquetting. In the present study common synthetic waste material i.e., different types of plastic, viz., Polyethylene (P.E.), Poly Vinyl Chloride (PVC) and Low Density Polyethylene (LDPE) is used as binder with molasses and lime. Briquettes were prepared and cold compressive strength (CCS) of these briquettes was estimated. From the current study it was observed that the among the new binders waste plastic i.e., PE seems to be very promising for the briquetting of chromites ore.

398539 Biochar Production and Application on Treatment of Flowback Water from Hydraulic Fracturing

Abstract:

Water shortage is a bottleneck for major shale plays in the U.S. and the world. Sustainable water reuse is urgent for the shale gas industry. In this paper, different biochar was prepared from wood chips by pyrolysis at different temperatures. The biochar prepared was tested on removal of inorganic salts, hydrocarbons, clay fine. The adsorption capacity of the biochar was also obtained for typical contaminants in flowback water from hydraulic fracturing industry.

401684 Production of Asphalt Binder from Ohio Coal Resources

Abstract:

The objective of this work is to develop a new coal end-use market by developing an advanced thermocatalytic technology, to convert coal into infrastructure materials including asphalt binder (binder), rejuvenator, and rheology modifiers with costs and performance equal to or better than those of petroleum-derived asphalt binders. Preliminary screening tests were completed using a 0.1 kg/hr solid feed fluidized-bed system. This system is well suited to evaluate the proposed process as it allows pulverized coal to be directly injected into a temperature controlled bed of fluidized material, providing excellent replication of commercial transport reactor operating conditions. Screening parameters such as Ohio coal type (Pittsburgh #8, Middle Kittaning, and Clarion No. 4A), operating temperature, feed gas composition, and addition of binder modifiers (styrene-butadiene-styrene (SBS), polyethylene (PE), and recycled crumb rubber (CRM)) were evaluated. Each sample was tested at operating temperatures ranging from 350-500°C. The results include interesting engineering data needed to refine an existing process techno-economic study and begin developing performance warranties for the binder product.

402856 Mercury Compliance While Avoiding Halogen-Driven Corrosion

Abstract:

EPA's Mercury and Air Toxic Standards (MATS) requiring stringent emission controls for coal-fired electric generating unit (EGU) power plants will be enforced beginning April 2015. Powdered activated carbons (PACs) have been demonstrated as the leading technology for achieving compliance with the stringent mercury control levels required by MATS. The PACs are sourced as halogenated or non-halogenated materials. In some cases halogens are added to the coal or flue gas stream separately from the PAC. As more of these systems are coming online and accumulating operating hours, issues of plant equipment corrosion are becoming known and work is being done to understand the nature and extent of the problems. Solutions can include operational changes and/or changes to the materials injected. We will discuss the design and application of advanced PACs to minimize or avoid corrosion issues and retain effective mercury capture and co-benefit integration with the EGU's air quality control system (AQCS).

401729 BACT in Air Pollution Control

Abstract:

Best Available Control Technology for air pollution is mandated by many regulatory agencies for projects above certain thresholds. The Prevention of Significant Deterioration (PSD) permitting guidance recommends selection of control technology using a top-down process. This session will cover the structure of the top-down selection process and basic cost effectiveness calculations using USEPA methods.

399709 Air Pollution Aftertreatment Control for Natural Gas Applications

Abstract:

Natural gas is an important part of the American infrastructure and economy, growing in significance with the rise of domestic shale gas. In 2013, 31% of the natural gas consumed in the United States was for electric power generation in 2,154 power plants. Regulated emissions from natural gas combustion include NOx, CO and VOC's (volatile organic compounds). These pollutants can be controlled by the use of a post-combustion catalyst installed downstream of the engine prior to the stack. For the power industry to provide electricity economically while maintaining low pollutant emissions, plant operators and equipment vendors face a variety of technological challenges which require innovative solutions.

NOx emissions are controlled by a Selective Catalytic Reduction (SCR) system. In a typical SCR system, gaseous ammonia (NH_3) is injected into the exhaust stream upstream of the SCR catalyst. In the presence of oxygen, NO and NO_2 react with NH_3 over the SCR catalyst to form N_2 and H_2O. Liquid ammonia or aqueous ammonia must be heated and injected into the unit in a precise amount in order to limit unreacted ammonia, also known as ammonia slip.

There are a number of issues currently facing the NOx control market for natural gas-fired applications. In order to improve air quality, regulators periodically reduce the emissions limits for new units, increasing the performance requirements of the technology, while users are trying to limit the cost and risks associated with operating these systems. Stricter limits require more catalyst and/or new designs from suppliers. As emissions limits become lower, the ammonia efficiency of the SCR system must be improved so ammonia injection is optimized and ammonia slip is minimized. Issues with ammonia quality are also common and can further increase operating costs by poisoning the catalyst and reducing the catalyst life cycle. As the emissions limits have been reduced, the emissions control system places more and more restrictions on the power producers and limits their operating flexibility to meet demand. To counter this, emissions control system designers are required to provide systems that operate outside the preferred operating window for the catalysts. Some large units require the catalyst to operate at higher temperatures. While on the other hand, faster start-up times and stricter start-up limits require the catalyst to operate at lower temperatures. Both of these issues are hardware and catalyst issues that require the catalyst supplier and system suppliers to work together to develop the right product.

Oxidation catalysts are used to oxidize CO and VOC's. The issues facing the oxidation catalyst are similar to the NOx control catalyst. The limits for CO and VOC's emissions are being reduced and operators are seeking more flexibility in terms of operating parameters. The big issue in terms of operating temperature has to do with poisoning of the catalyst by sulfur. At lower temperatures sulfur blocks the active sites of the oxidation catalyst limiting activity. At higher temperatures the sulfur desorbs, which is why it is a temperature-dependent issue. There are methods to remove the sulfur from the catalyst, but as long as it is present in the fuel the issue will return. Finding formulations that operate at lower temperatures is an important area of development currently.

Solar Energy Harvesting and Storage Using a Salinity Gradient Solar Pond Under the Northern Cyprus Climate Conditions-Experimental Investigations and CFD Simulation

[1]Soudabeh Gorjinezhad, [2]Mohammad Askari, [1]Negar Zare Pakzad, [1]Dilem Eroğlu, [1]Abdul Rehman Rajab Habib, [1]Sevgi Gödelek, [1]Mehdi Amouei Torkmahalleh, [1]Orhun Kahraman, [1]Sultan Kadyrov, [3]Goodarz Ahmadi

1- *Chemical Engineering Program, Middle East Technical University Northern Cyprus Campus, Guzelyurt, Mersin 10, Turkey*
2- *Mechanical Engineering Program, Middle East Technical University Northern Cyprus Campus, Guzelyurt, Mersin 10, Turkey*
3- *Mechanical Engineering Department, Clarkson University, Potsdam, NY, 13699- 5725*

1. INTRODUCTION

Today, renewable energy sources gain importance day by day. It is crucial to develop devices and processes to supply energy from non-polluting and renewable energy sources for sustainable development of the world. Solar pond is an example of such devices that basically collects solar energy and stores it as thermal energy for a long period. Studies indicate that the temperature in solar ponds may in general reach up to 70-80°C implying that the thermal energy from solar ponds can be useful to applications with low grade energy demand [1]. One of the most important advantages of solar pond over other renewable energy sources, such as solar collectors, is its lower cost of investment [1]. Solar pond is environmentally friendly in particular when it is used for electricity generation. The heat obtained from solar pond can be converted into electric power even at low temperatures [2]. In this regard, organic Rankine cycle engines are generally run using the temperature difference in a solar pond. For applications where the organic fluid fails to operate due to low temperature difference, thermoelectric generators can be a good candidate to replace organic Rankine cycle engines for power generation [3].

Solar Ponds normally consist of three different salinity layers. The first layer, known as the upper convective zone (UCZ), is located at the top of the pond, and contains the least salinity level. The second layer, whose salinity level increases with depth, is called non-convective zone (NCZ). This layer is responsible to act as an insulator to prevent heat from escaping to the UCZ, maintaining higher temperature at deeper zones. The last layer made of a saturated salt solution, is responsible for energy storage, and is known as the lower convective zone (LCZ) [5]. The working principles of solar ponds are quite simple. In the absence of salinity gradient, the solar radiations reaching a pond are mostly absorbed by the lower water levels and will cause the water to heat up. The heated water is then risen up to the pond surface due to buoyancy effect and loses its thermal energy to the atmosphere. Thus, the main reason behind imposing salinity gradient to solar ponds is to create a density gradient restricting such buoyancy driven natural convection and as a result to trap the thermal energy at the bottom of the pond [4]. The performance of the solar pond may decrease with the increasing evaporation rate, and decreasing the salinity gradient [6]. Thus, it is critical to prevent these phenomena.

There have been several modeling and simulation studies carried out on salinity gradient solar ponds, and were validated by experimental data. Akbarzadeh et. al. [2] performed some experimental studies to validate their newly developed model. The model then was used to evaluate the performance of a large industrial solar pond. They employed an overall and a layer-wise energy balance in their model, taking into account all possible modes of heat transfer to evaluate the heat transfer rates in each zone of the pond. The theoretically predicted temperature value indicated higher variations from the experimental values at the UCZ compared to other layers, mostly because of more significant environmental effects on this zone. Mehdizadeh and Ahmadi [4] studied the stability, temperature evolution, and salt diffusion in a solar pond using a two-dimensional unsteady CFD model in ANSYS FLUENT. Their results signified the importance of regular concentrated brine injection to accommodate for salt diffusion and maintain the stability and long-term effectiveness of the pond. Jafarzadeh [5] applied the finite difference method to study thermal behavior of a large salinity-gradient solar pond in the city of Mashhad, Iran, based on various layer thicknesses. His findings contribute to determining the optimum thickness for upper, middle and lower zones. In addition, it has been reported that the maximum heat removal from the pond can be at most about 15% or 22% for constant and variable loading, respectively. Ouni et. al. [6] developed a one dimensional model to simulate the transient behavior of a pond located in Tunisia. The simulation results suggested that the maximum heat that can be extracted during summer season is 80 w/m^2, while in winter the rate of heat removal should be reduced by 75 per cent, in order to maintain an appropriate temperature in LCZ. Shojaeefard et. al. [7] performed a computer simulation in COMSOL MULTIPHYSICS considering all thermal and hydrodynamic properties of the solution to be varying by temperature and validated their results by comparing to the experimental data of Ouni et. al. study [6]. In their study, the model exhibited insignificant change in temperature and concentration profiles in any direction other than with depth. Sakhireh and Al-Salaymeh [8] investigated the performance of a solar pond under Jordanian climate conditions and utilized the mathematical MATLAB model developed by Jafarzadeh [9]. The result obtained from the numerical model indicated that temperature at LCZ is directly proportional to ambient temperature where in Jordanian climate a maximum temperature of 47°C is achievable. Giestas et. al. [10] carried out a two-dimensional numerical study on a solar pond with density and viscosity dependence on temperature and salt concentration. Their findings justified the significant influence of variable viscosity dependence model, especially in assessing thermal losses at the surface of a solar pond. Suarez et, al. [11] developed, for the first time, a fully coupled two-dimensional numerical double-diffusive convection model to examine temperature behavior and stability of the pond.

Northern Cyprus is enriched in solar energy in particular during summer. The performance of a solar pond in this region has not been evaluated yet. To take advantage of solar energy in Northern Cyprus, a salinity gradient solar pond has been constructed and operated at Middle East Technical University Northern Cyprus Campus (METU NCC) located at Guzelyurt, Northern Cyprus (***Figure 1***). This study is primarily concerned with describing the development, validation and use of CFD for computer modeling of flow and thermal analysis of the experimental salinity gradient solar pond at METU NCC. The pond has started operating since Oct. 2014 and experimental results are being recorded accordingly. The CFD commercial software considered to be utilized is ANSYS FLUENT and the challenge is to visualize the transient effect of temperature distribution, salinity concentration, velocity field and the degree of influence of wind speed on the overall stability and performance of the solar pond over time. Furthermore, the model enables us to evaluate the performance of a larger scale solar at METU NCC.

Figure 1. Constructed Solar Pond at Middle East Technical University (Northern Cyprus)

2. MATERIALS AND EXPERIMENT

A cylindrical salinity gradient solar pond of 61 cm diameter, 55 cm height and 1.2 cm thickness installed and operated since October 8th, 2014 at Middle East Technical University Northern Cyprus Campus (METUNCC) located at Guzelyurt, Northern Cyprus. Three inlet ports were installed to carefully add the solutions to the pond to create the three salinity layers. Also, these inlets were used to add proper amount of solutions to each zone during the experiments to compensate for evaporation and sampling losses as well as surface washing. The pond has been divided in three zones. The bottom zone (LCZ) consisted of 75 liters of saturated salt solution (concentration C). The middle zone (NCZ) consisted of three equally divided sub-layers, each 15 liters in volume, with 0.75C, 0.5C and 0.25C concentrations. The top zone (UCZ) consisted of 15 liters of fresh water. The rest of the pond was left empty. The pond was equipped with three sampling valves to withdraw samples from each layer to monitor the salt concentrations using conductivity measurements. Six thermometers (2 in each zone) were installed to monitor the temperature variations at 9 a.m., 1 p.m., 5 p.m., and 10 p.m. every day. Temperature recording at other time of the day was performed as needed. The bottom and sides of the pond were insulated with a 2.5cm thick thermal insulator. The inner surfaces of the pond were painted black. Excess (undissolved) salt was added to the bottom of the pond, to ensure saturation of the bottom layer. The ambient temperature and solar radiation were monitored using a sun tracker (Kipp and Zonen) placed next to the solar pond. Wind velocity and relative humidity were also monitored throughout the experiment.

3. COMPUTATIONAL SCHEME AND MODELING

The geometry and meshing of the solar pond were constructed by ANSYS Workbench Design Modeler and Meshing environment. FLUENT version 14.5 was used as the CFD solver. In accordance with most salinity gradient solar pond studies, a two-dimensional computational domain is considered for investigating the transient behavior of the pond representing a cross-section of the solar pond. Accordingly, the rectangular domain of consideration is that of the existing pond dimensions with 61 (cm) in width and 50 (cm) in height. The lower wall of the rectangle is assumed to be the ground and upper boundary to be the top water level open to atmosphere. Vertical walls and the bottom surface of the pond are set to be impermeable and thermally insulated with the corresponding thermal conductivity value of the insulation material. Also the top surface, due to contact with ambient air, is taken to be at ambient temperature with an average wind speed according to local meteorological data. The salinity concentration of the LCZ is kept at saturated level, while the top surface is set to zero concentration, initially.

The modeling of the pond is divided into two phases. Firstly, a simple model is going to be developed in which it is assumed that the concentration level remains the same due to the present regular injection of brine solution to the system. Accordingly, the thermofluid properties will be assumed to be primarily a function of temperature and will be evaluated for each layer at its initial concentration. Thus, in the first phase, the concentration will not be considered as a modeling parameter and it is aimed to primarily study the variation of the temperature using the most simplified model possible. In the second phase, the model will be enhanced by defining the concentration as a varying parameter to the system. In this regard, the variation of properties with concentration can be taken into account and the significance of salt diffusion and its influence on the stability of the pond can be investigated. In addition, the results between both cases and the experimental results will be compared to justify the computational complexity added to the model. The following modifications is necessary to be done for modeling of the pond in FLUENT.

3.1 Mass Transfer Equation for Salt Diffusivity

The concentration of the salt should be defined as a UDS (User Defined Scalar) to FLUENT. In a simplified manner, Fick's law can be employed to model the sodium chloride diffusion within the pond. Since the thermal diffusion in solar ponds are reported to be about only 10 percent of the total diffusion, the convective terms of the mass transfer scalar transport equation can be ignored. The rate at which the particles are generated or dissipated per unit volume can also be neglected from the governing equation, resulting into the following form for two dimensional domain [7].

$$\frac{\partial C}{\partial t} = \frac{\partial}{\partial x}\left(D\frac{\partial C}{\partial x}\right) + \frac{\partial}{\partial y}\left(D\frac{\partial C}{\partial y}\right) \qquad (1)$$

3.2 Thermal and Hydrodynamic Fluid Properties Model

The main challenge in the present study, is that all the hydrodynamic and thermal properties of the sodium chloride solution are considered to be varying both with temperature and concentration. As a result, energy and mass transfer equations need to be coupled and therefore, a difficulty at the modeling stage as well as iterative solution solving arises. If the parameters existing in energy equation did not depend on concentration, it would be possible to solve for the temperature field initially and accordingly obtain the concentration distribution from the already known temperature field [6].

To account for the variation of density with respect to temperature as well as concentration, a modified version of Boussinesq Approximation was used in the present study. FLUENT accounts for the variation of density with temperature through Boussinesq model; yet, the variation with concentration needs to be added with the use of a UDF (User Defined Function). β_t and β_m in the following equation are the thermal and mass expansion coefficients.

$$\rho(T,C) = \rho_0 + \left(\frac{\partial \rho}{\partial T}\right)_P (T - T_0) + \left(\frac{\partial \rho}{\partial C}\right)_C (C - C_0) = \rho_0\big(1 + \beta_t(T - T_0) + \beta_m(C - C_0)\big) \qquad (2)$$

The variation of thermal conductivity can be approximated as a linear function of temperature and concentration in the following form [12].

$$k\,(T,C) = a_1 + a_2(T - 20) + a_3 C \qquad (3)$$

$$Where\ a_1 = 0.5553\ ,\ a_2 = 8 \times 10^{-3}\ ,\ a_3 = 8.13 \times 10^{-5}$$

Specific heat of the fluid at constant pressure can be considered to be a function of temperature and concentration according to the following equation [13]. The term ‘s’ in the following equation refers to the salinity of the solution and is simply the concentration divided by the solution density.

$$C_p\,(T,s) = b_1 + b_2\,T + b_3\,T^2 + b_4\,T^3 \qquad (4)$$

$$s = \frac{C}{\rho}\left(\frac{kg\ salt}{kg\ solution}\right)$$

$$Where\ b_1 = 4.2068 \times 10^3 - 6.6198 \times 10^3\,s + 1.2288 \times 10^4\,s^2$$

$$b_2 = -1.1262 + 5.4178 \times 10\,s - 2.2719 \times 10^2\,s^2$$

$$b_3 = 1.2026 \times 10^{-2} - 5.5366 \times 10^7\,s + 1.8906\,s^2$$

$$b_4 = 6.8774 \times 10^{-7} + 1.517 \times 10^{-3}\,s - 4.4268 \times 10^{-3}\,s^2$$

The following equation accounts for the variation of the molecular diffusion coefficient for a temperature range of 5°C to 100°C and salinity level of 0 to 20 percent with reasonable accuracy [14].

$$D(T,s) = (c_1 + c_2\,T + c_3\,T^2 + c_4\,s + c_5\,s^2) \times 10^{-10} \tag{5}$$

$$Where\ c_1 = 8.16\ ,\ c_2 = 0.255\ ,\ c_3 = 2.54 \times 10^{-3}\ ,\ c_4 = -0.28\ ,\ c_5 = 1.47 \times 10^{-2}$$

3.3 Solar Energy Model, Heat Source Term

The solar energy absorption within the pond is modelled using the Beer-Lambert law of solar energy absorption. According to this law, it is known that most of the solar radiation striking the pond is absorbed in rather upper layers of water. The absorption coefficient of the solution, as for any other substance, varies with the wavelength of the incident light. In FLUENT, a User Defined Function is to be compiled to evaluate the resulting volumetric heat source term at each time step for based on the elevation of the cell from the top surface [15]. In the following equation, q_0 represents the solar irradiation reaching the surface of the pond. A sinusoidal equation can be written to approximate the variation of penetrated solar irradiation onto the surface, q_0, as a function of time per day. The constants γ_i and ϵ_i which depend on the absorption of different wavelengths are also given. Therefore, the energy source term, S_R, is the derivative of q_z and is given in Equation (7).

$$q(Z) = 0.85\,q_0 \sum_{i=1}^{4} \gamma_i \exp\left(-\frac{\epsilon_i H}{0.76}\,(1 - Z)\right) \tag{6}$$

$$S_R = \frac{\partial q(Z)}{\partial z} = 0.85\,q_0 \sum_{j=1}^{4} \frac{\gamma_i\,\epsilon_j}{0.76} \exp\left(-\frac{\epsilon_j}{0.76}\,(H - Z)\right) \tag{6}$$

Wavelength	γ	ϵ
0.2-0.6	0.237	0.032
0.6-0.75	0.193	0.45
0.75-0.9	0.167	3
0.9-1.2	0.179	35

4. Results and Discussion

Figure 1 shows the average monthly solar pond temperature variations from October 2014 to February 2015). Figure 5 also presents average monthly ambient temperature as well as irradiance. As can be seen, the pond has been functioning as the temperature differences among the three layers have been established. The pond temperature variations are well correlated with the ambient temperature variations. The temperature at all layers decreased from October 2014 to end of December 2015, but then beginning January 2015, it increased as solar irradiance increased.

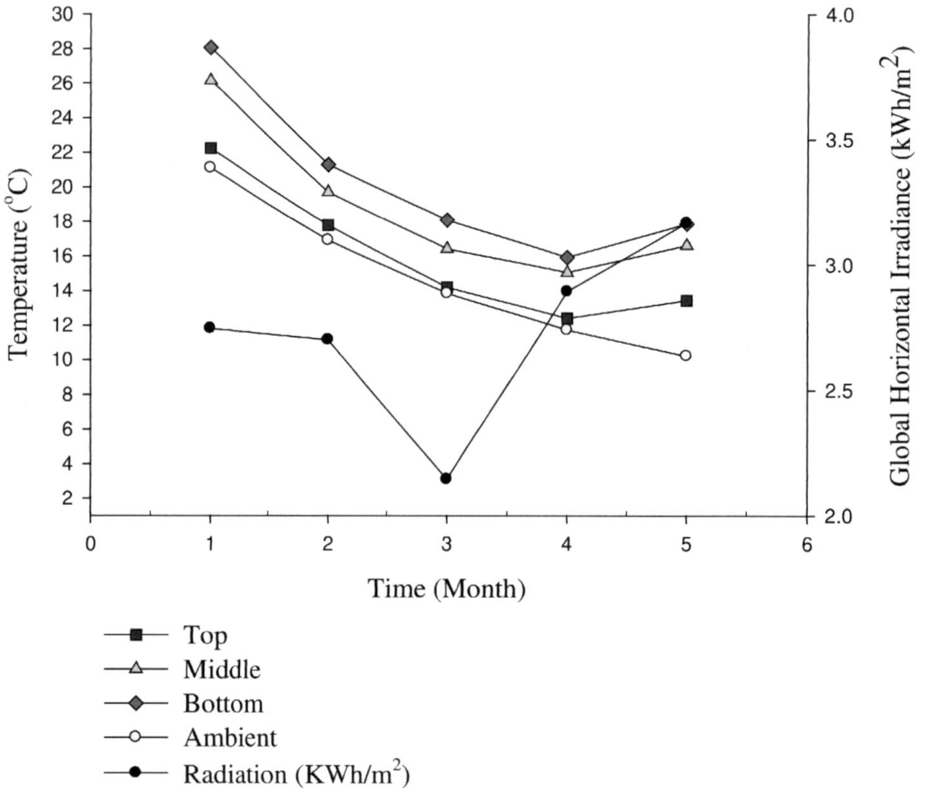

Figure 1. Average monthly temperature variations of the solar pond

REFERENCES

[1.] Sukhatme, S., & Nayak, J. (2008). *Solar energy: Principles of thermal collection and storage* (3rd ed.). New Delhi: Tata McGraw-Hill.

[2.] Akbarzadeh, A., Bernad, F., Casas, S., Gibert, O., Cortina, J. L., & Valderrama, C. (2013). Salinity gradient solar pond: Validation and simulation model. *Solar Energy*, *98*, 366-374.

[3.] Singh, B., Gomes, J., Tan, L., Date, A., & Akbarzadeh, A. (2012). Small Scale Power Generation using Low Grade Heat from Solar Pond. *Procedia Engineering,* 50-56

[4.] Mehdizadeh, M., & Ahmadi, G. (2014). Two-dimensional Computer Simulation of Salinity Gradient Solar Pond Operation. *Proceedings of ASME 2014 Fluids Engineering Summer Meeting, FEDSM2014.*

[5.] Jaefarzadeh, M. (2005). Thermal behavior of a large salinity-gradient solar pond in the city of mashhad. *Salinity Gradient Solar Pond.* Retrieved October 14, 2014, from http://profdoc.um.ac.ir/articles/a/203017.pdf

[6.] Ouni, M., Guizani, A., & Belguith, A. (1998). Simulation of the transient behaviour of a salt gradient solar pond in Tunisia. *Renewable Energy*, *14*(1-4), 69-76.

[7.] Shojaeefard, M.H., Tafazzoli Aghvami, K., & Mazidi, M. (2013). Computer Simulation of Performance and Transient Behavior of a Salt Gradient Solar Pond. *JöKULL Journal*, *63*, 58-76.

[8.] Sakhrieh, A., & Al-Salaymeh, A. (2013). Experimental and numerical investigations of salt gradient solar pond under Jordanian climate conditions. *Energy Conversion and Management*, 725-728.

[9.] Jaefarzadeh, M. (2004). Thermal behavior of a small salinity-gradient solar pond with wall shading effect. *Solar Energy*, 281-290.

[10.] Giestas, M., Pina, H., Milhazes, J., & Tavares, C. (2009). Solar pond modeling with density and viscosity dependent on temperature and salinity. *International Journal of Heat and Mass Transfer*, 2849-2857.

[11.] Suárez, F., Tyler, S., & Childress, A. (2010). A fully coupled, transient double-diffusive convective model for salt-gradient solar ponds. *International Journal of Heat and Mass Transfer*, 1718-1730.

[12.] Karakilcik, M., Kiymac, K., & Dincer, I. (2006). Experimental and theoretical temperature distributions in a solar pond. *International Journal of Heat and Mass Transfer*, 825-835.

[13.] Patankar, S. V. (1980). Numerical Heat Transfer and Fluid Flow. *Taylor & Francis*.

[14.] Angeli, C., & Leonardi, E. (2005). The effect of thermo-diffusion on the stability of a salt gradient solar Pond. *International Journal of Heat and Mass Transfer*, 4633-4639.

[15.] Boudhiaf, R., Ben Moussa, A., & Baccar, M. (2012). A Two-Dimensional Numerical Study of Hydrodynamic, Heat and Mass Transfer and Stability in a Salt Gradient Solar Pond. *Energies Journal*, 3986-4007.

401169 Time-Dependent Multi-Period Assessment of Life Cycle Cost and Environmental Impacts of CO2 EOR

Abstract:

EOR (enhanced oil recovery) methods has been widely adopted for increasing oil production rate, and injection of CO_2 into the reservoir other than other mediums in EOR applications can be favorable in these days, due to CCS (carbon capture and sequestration). Although sustainable gains can be obtained from CO_2 EOR, considerable investment and energy consumptions are needed to accommodate CO_2 purchase, CO_2 injection, product separation and CO_2 recovery. Therefore, it is helpful to perform LCA (life-cycle analysis) for gathering an inventory of relevant energy and material inputs and environmental releases related to CO_2 EOR operation, and evaluating them in a systematic and integrated manner.

A few studies conducted for life cycle analysis of CO_2 EOR have been mainly focused on compiling overall energy and material inputs for whole life time of EOR, and assessing total amount of CO_2 purchase, CO_2 recycle and oil production as a whole. These lumped life-time values may not be adequate for interpreting time-dependent and non-constant characteristics in EOR operations, for example, design and operation of surface facilities in CO_2 EOR, . Hence, in this study, multi-period approach considering time-dependent characteristics of CO_2 EOR operation is introduced for LCA. Process modeling for CO_2 EOR production facility is coupled with LCA, with which processing capacity and energy consumption can be simultaneously considered. Also multi-period consideration of material and energy inputs allows detailed interpretation of life time impacts of CO_2 EOR. A case study is presented to highlight the importance of time-dependent consideration on CO_2 EOR operation and to effectively quantify key economic and environmental trade-offs through LCA.

Acknowledgment

This work was supported by “The development of CO_2 geological storage technologies through 1,000 ton CO_2-EOR pilot test” of the Korea Institute of Energy Technology Evaluation and Planning (KETEP) grant funded by the Ministry of Trade, Industry and Energy (MOTIE) (No.2012T100201728).

401765 Improving Process Sustainability for a Large U.S. Gray Iron Foundry

Abstract:

Improving Process Sustainabililty for a Large U.S. Gray Iron Foundry

Prashant Nagapurkar, Shyam Paudel and Joseph D. Smith, PhD.

Missouri University of Science and Technology, Rolla, Missouri, USA

Abstract

Sustainability has been a key issue in many industrial processes. In 2012, the U.S. produced 12 million tonnes of metal castings equivalent to $34 billion in total revenue. The metal casting industry will continue to face unstable energy markets as well as stringent Environmental Protection Agency (EPA) regulations for their greenhouse gas emissions. Therefore, it is necessary to evaluate the environmental effects of the foundry process. This study conducts a life cycle analysis for assessing environmental impacts associated with the production of gray iron castings at the Caterpillar^{TM} foundry at Mapleton, IL. Initially, a base case study has been developed on Powersim^{TM} software to estimate the CO_2 emissions for the existing process at Mapleton foundry. Based on our plant visit and its process assessment, we have proposed advanced control strategies for energy intensive variables like melters, fans, dust collectors and burners to optimize and reduce energy consumption. After implementation of these control strategies our simulation results show that Mapleton foundry has a potential to reduce their CO_2 emissions by 18% when compared to the base model thereby making their process more sustainable and profitable.

397207 Influence of Ice Cavity Geometry Changes on Burning Behavior of Large Scale Crude Oil Fires

Abstract:

Several Arctic council reports conclude that oil spills are the most significant threat to the Arctic ecosystem. Some studies have shown that *in-situ* burning (ISB) on open water can remove more than 90 % of the spilled oil, and it is the most promising technology for an efficient response to oil spills in Arctic. However, few studies have been undertaken on the influence of ice on the efficiency of oil burn. Prior studies have shown the coupling between the mass loss rate () and the geometry change of the ice cavity in pool fires smaller than 25 cm [1, 2]. In this study, a series of experiments are performed to develop an understanding of the influence of the coupling of ice cavity geometry changes and burning behavior of larger sized pool fires. The experiment involves Alaska North Slope (ANS) crude oil ignited in ice cavities of diameters (25-100 cm) with depth 6-25 cm. It is found that both burning rate and radiative heat flux are enhanced because of the cavity expansion. However, the overall efficiency of the burn is significantly lowered due to formation of lateral cavities. At some stage during the burning process, the heated oil tends to penetrate into the ice forming a pocket. This lateral cavity causes 10 - 35% of the crude oil to be trapped inside and hence un-recoverable. Further understanding of the reasons that may cause this unique formation and measures to prevent the same are discussed.

References

[1] P.W. Bellino, A.S. Rangwala, M.R. Flynn, A study of *in situ*burning of crude oil in an ice channel, Proceedings of the Combustion Institute, 34 (2013) 2539-2546.

[2] H.F. Farahani, X. Shi, A. Simeoni, A.S. Rangwala, A study on burning of crude oil in ice cavities, Proc. Combust. Inst., 35 (2014).

This study was funded by the Bureau of Safety and Environmental Enforcement, US Department of the Interior, Washington, D.C., under Contract Number E12PC00056. The contents do not necessarily reflect the views and policies of the BSEE, nor does mention of the trade names or commercial products constitute endorsement or recommendation for use.

CFD Analysis of Supercritical Water Reactor for Flow Back and Produced Water Treatment

Swapnil Fegade, Abhijeet Shrawage, Jason Trembly

Ohio Coal Research Center

Ohio University, Athens OH 45701

Shale deposits have the potential of unlocking a vast reserve of fuel across the United States. However, the current high water usage of the horizontal drilling and hydraulic fracturing techniques used to develop shale resources needs to be addressed. OHIO is developing the Integrated Precipitative Supercritical Process (IPSC), with the aim of reusing flowback and produced water (F/P) from shale wells to reduce the water intensity of shale development activities. Of the components in the IPSC process, the Supercritical Water Reactor (SCWR) will be responsible in the treatment of the major salt constituents in F/P water. The purpose of this work was to utilize CFD analysis, using ANSYS Fluent (Fluent), to investigate the heat transfer and fluid flow in the SCWR, at different inlet temperatures (350° C and 390° C) and operating pressures (23 MPa and 26 MPa).

A user-defined material representing Supercritical Water (SCW) was developed in Fluent, which was later compared to existing experimental data from literature and implemented to evaluate SCWR operation. Temperature and density contours along with flow patterns within the SCWR are presented. The estimated TDS solubility in the lower half of the SCWR were estimated based upon SCW density. It was observed that at higher operating pressures, SCW could be heated to a higher temperature compared to pressures around the supercritical point (PSupercritical = 22.064 MPa). However, the minimum SCW density increases as the operating pressure increases, resulting in an increase of TDS solubility. As a result, the ideal conditions would need to be a balance between the inlet temperature of F/P water and the operating pressure of the system.

400845 Arsenic Removal from Aqueous Solution Using Nano-Structured Calcium Silicate

Abstract:

In this study, removal of arsenic from water was intended with using nano-structured calcium silicate as an adsorbent which was produced from rice hull ash.

For calcium silicate, two different production methods, which are calcium silicate produced in the presence of water and spacer (diethylene glycol), were performed. Arsenic removal efficiencies were compared and nano-structured calcium silicate which was produced in the presence of spacer compound was choosen according to the experimental arsenic removal results.

The study covers the determination of the effect of time, pH, different arsenic concentrations and adsorbent quantities to obtain the optimum conditions for arsenic removal with nano-structured calcium silicate.

401322 Optimal Strategies for NOx Reduction in Power Generation Using Optimization Approach

Abstract:

The power generation sector is considered as one of the major contributors to air pollution. An optimization model was formulated and written in a general format. The objective of this model is to select the best pollution control strategy for the power generation to reduce NO_x to specific level while meeting the electricity demand at minimum cost. The model is illustrated in a case study taken from Ontario Power Generation; Canada. For the NOx case when we consider two options (fuel balancing and switching), it was found that fuel switching is the best option of choice for reduction targets from 5 up to 30%.However, by considering all mitigation options, the result shows that applying SCR technology is the best option to reduce NOx emissions.

Tier 3 Vehicle Emissions and Fuel Standards Compliance - A Sustainability Perspective

Jacqueline Sibblies, PE
Environmental Compliance Consulting
P.O. Box 5013
Somerset, NJ 08873
sibbliesj@gmail.com
(908) 531-4029

INTRODUCTION

By the 1980's, global awareness of the detriment of air pollution to ecological systems started escalating. Several industrialized western nations, including the United States, began recognizing the intricate correlation between air and water pollution, and the cumulative negative effects on plant and animal life, both aquatic and terrestrial. This knowledge led to legislative action, resulting in the promulgation of initiatives/directives on and/or regulations of materials, processes and industrial operations to bring about significant reductions in the rates of pollutant emissions to the atmosphere.

In the United States, Congress authorized the Clean Air Act Amendment of 1990, which was comprised of a comprehensive (and at the time a seemingly stringent) approach to attaining air pollution control from sources identified as significant contributors of air pollution, particularly criteria pollutants (VOC, NOx, SOx, CO, PM and Pb) and hazardous air pollutants (HAPs). Many of the sources of these emissions were fuel combustion sources.

Specifically to achieve reduction in emissions of sulfur oxides (SOx) and lead (Pb), the sulfur and lead contents of petroleum products, like diesel and gasoline, had to be reduced. This placed a heavy demand on industry to develop and implement the kind of technology that could accomplish these regulatory objectives. And industry rose to the occasion.

The petroleum refining, automobile and supporting industries worked to attain performances from combustion engines and systems that met and sometimes exceeded the regulatory standards. This new era made terms like "unleaded fuel" and "low-sulfur fuel" common consumer lexicons.

Continuous monitoring of the environment has revealed the need for continuous improvement. Environmental regulators globally determined a need for further reduction of air emissions from petroleum combustion sources and consequently further reduction of sulfur content in fuel. This is evident by similar, simultaneous regulatory actions world-wide - from international organizations like United Nations Environment Program (UNEP) and International Council on Clean Transportation (ICCT) working on a program to develop a global sulfur strategy, to the European Union's (EU) effecting directives on stringent sulfur-in-fuel standards (Euro 5) to

China doing similarly in "China 5", to the United States' promulgation of the Tier 3 Fuel Standards and Vehicle Emissions Program.

Fortuitously, the increasing need for improvements in environmental quality resulting in more stringent regulatory restrictions on fuel content and improved motor vehicle performance is contemporaneous with the explosive advancement in information technology (IT) since the turn of the 21st Century. This growth has expanded into advancements of several industry sectors, like communication, healthcare, transportation and surveillance. IT is applicable to all industries, so industries subject to these regulatory requirements may take advantage of this resource and the opportunity to collaborate for advancement, which will be profitable for business while improving the environment and enhancing human health. This paper will explore the sustainability benefits of the Tier 3 Program and like regulatory standards.

THE TIER 3 PROGRAM

The Tier 3 Fuel Standards and Vehicle Emissions Program (Tier 3 Program) was promulgated by the United States Environmental Protection Agency (USEPA) and applies to several classes of vehicles. This program, which comes into effect on January 1, 2017, employs a comprehensive approach to further restrict the primary pollutants that detriment ambient air quality. It builds on the Tier 2 Program, promulgated in 2000, in which the vehicle and fuel were treated as a system in order to gain optimal results from the emission reduction requirements.

Table 1. Standards for Light Duty Motor Vehicles

STANDARDS Emissions	USEPA Tier2 (Current)	USEPA Tier3 (2017-2025)	California CARB LEV II/III	China (2018)	EU Euro V/VI
$NMOG+NO_x$	160 mg/mi	30 mg/mi	0.16 g/mi	0.230 g/Km	128 mg/Km
PM	10 mg/mi	3 mg/mi	0.01 g/mi	4.5 g/Km	5.045 mg/Km
Fuel Vapor		0.300 g/test	–	–	–
Sulfur-in-Fuel					
Gasoline	30 ppm	10 ppm	10 ppm	10 ppm	10 ppm

Values are fleet averages measured on the Federal Rest Procedure (FTP)
CARB – California Air Resources Board
mg/mi – milligrams per mile; g/test – grams per test; ppm – parts per million; g/Km – grams per kilometer

As indicated in Table 1, the Tier 3 standards include further restrictions on motor vehicle tailpipe emissions such as non-methane organic gases (NMOG), nitrogen oxides (NO_x), particulate matter (PM) and evaporative emissions, specifically fuel vapors. The program also sets a much lower limit for fuel sulfur content, 60% less than the Tier 2 standard, but comparable to other existing requirements across national and international territories. In effect, the Tier 3 program has aligned with existing programs across the globe

MOTOR VEHICLE AND THE ENVIRONMENT

In an ideal scenario, petroleum fuels like gasoline and diesel when burned completely, yields only energy, carbon dioxide (CO_2) and water (H_2O). However, real-life situations are far from ideal. Gasoline and diesel are not pure hydrocarbons, but contain substances other than hydrocarbons, like sulfur and lead which also react in the combustion process, producing sulfur and lead compounds.

These and other byproducts of combustion have created hazardous environmental conditions. Several western nations including the United States started conducting environmental assessments in the mid 20th Century in an effort to determine the sources of environmental pollutants that were threatening several sectors of the ecosystem. Smog, for example, was a major concern d in major cities; it was visible and visibly worsening. Environmental assessments conducted by national agencies revealed that petroleum fuel combustion was a major contributor to smog and other environmental hazard conditions, and though industrial processes constituted a large contributing factor, so too was motor vehicle operation.

In the 1950's and 1960's, regulatory requirements were put in place in an effort to control environmental pollution, including from motor vehicles. Increased knowledge of pollution sources, their impacts on the environment and the need for a comprehensive approach for controlling the emissions and mitigating their effects eventually led to the creation of statutory agencies authorized to manage environmental programs. The USEPA is one such agency, established in 1970 to implement environmental regulatory programs including the new Clean Air Act, which mandated reduction in motor vehicle tailpipe emissions of up to 90%.

Regulatory Impacts

This emission reduction requirement necessitated that motor vehicle manufacturers and emissions control equipment manufacturers (closely associated industry) collaborate on a technological solution. By the mid 1970's, the industries developed the catalytic converter which were being assembled to the new model vehicles to meet the emission compliance of the Clean Air Act.

Although the catalytic converter performed well in reducing tailpipe pollution (via oxidation/reduction reactions of the fuel combustion by-product gases before they reach the atmosphere), it was discovered that lead from the fuel undermined the performance of the equipment and thus its useful life. This discovery, in addition to the recognition of the severe health hazard of lead in the atmosphere, led regulations on the reduction of lead in gasoline. And so the contribution of petroleum industry was incorporated in the evolving correlation of industrial technological development and regulatory compliance for environmental improvement

THE PETROLEUM INDUSTRY

In the Unites States, lead in leaded gasoline was phase out in the 1980's. During this decade, society had become aware of another element in fuels that was causing adverse effects on the ecosystem – sulfur. Sulfur in fuel created combustion by-products of sulfur oxides (SOx) which dissolved in precipitation to form acids, creating very unfavorable conditions for land and aquatic life.

National environmental agencies initiated legislation and directives, requiring that the petroleum industry provide fuel with reduced sulfur content. Petroleum refineries, in particular, had to develop technological changes to reduce sulfur (which is a natural component of crude oil – up to 0.5%) from marketed fuel.

Regulatory requirements to reduce sulfur content in fuel have occurred in phases over the last twenty-five (25) years, so not only is the petroleum industry familiar with requisite technology for accomplishing reduction, but the industry should have experienced benefits from reduced sulfur content in its operating systems.

SUSTAINABILITY OPPORTUNITIES

Sulfur compounds do not only produce acids that are detrimental to the ecosystem, they also form acids during the petroleum refining and processing operations, in the equipment and piping. Acids can damage equipment and affect their useful life. The resources required by the industry to meet compliance with Tier 3, like modifying operations and/or retrofitting equipment operations may also be applied to assessing parallel changes that would improve performance. This could be further reduction in fuel sulfur content and/or other elements affecting the quality of fuel. The industry and affiliate industries like motor vehicle manufacturing would benefit, and so would the environment and society, adding a sustainable dimension to an environmental compliance project.

Applying sustainability principles in plans for significant change serves to optimize results because a holistic approach is applied. This approach is comprehensive; all affected areas of operation are considered and possibilities for change explored.

It is apparent that requirements on sulfur content in fuel will become more stringent, getting very close to zero. The state of California, being the established front runner (nationally and thus internationally) on mitigating air pollution, has already indicated this progression with the introduction of the LEV III program. Requirements include standards for ultra low sulfur-in-fuel and zero emission vehicles (ZEV). The standards may be deemed necessary, considering population growth, the steady increase in the number of motor vehicles on the roads, and the spread of industrialization globally. All of these activities would result in increased pollution – all other factors remaining unchanged. This being the case, environmental stakeholders need to plan ahead for ways and means to control a growing problem.

As shown in Figure 1. below, SOx pollution became an acute problem in the 1970's, after beginning a sharp increase in the late 1940's after World War II, and the beginning if the baby boom generation. Emissions steadily decreased with the implementation of environmental programs that limited releases to the atmosphere.

Figure 1.

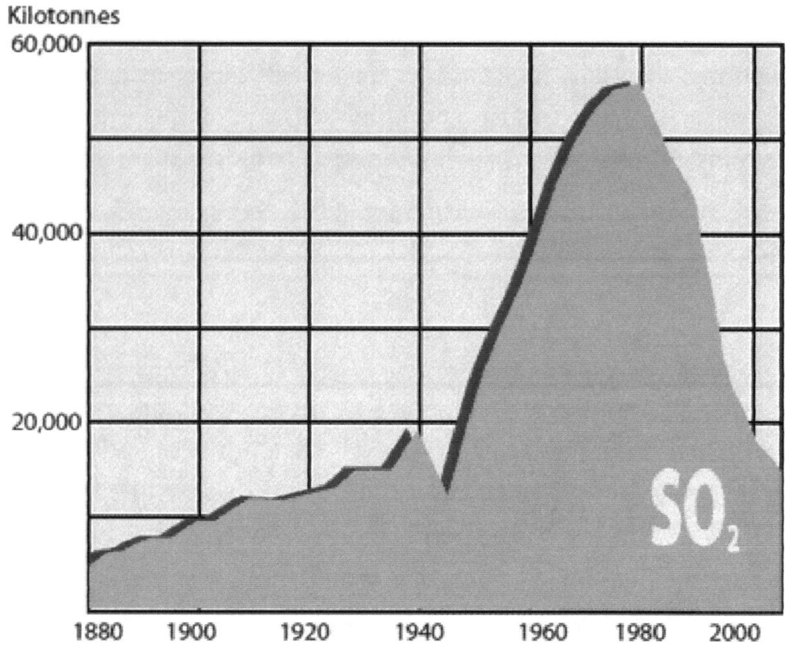

Figure: Emissions of sulphur dioxide in Europe over the period 1880-2005.
Source: Vestreng et al, Atmospheric Chemistry and Physics, 2007

The motor vehicle manufacturing industry could also benefit from applying a sustainability approach planning for compliance with Tier 3. The combustion process that occurs in a motor vehicle engine is far from being ideal. It is estimated that just over 40 percent of the fuel fed into a vehicle's engine is converted into work, i.e. sustaining the functioning of the vehicle (the average varies for the types of vehicle). More than 50% of the fuel fed is dissipated as heat. Planning for compliance of Tier III provides an opportunity to apply principle of sustainability, and evaluate current technology for areas of performance improvement. A more efficient engine means less air pollution, and thus a cleaner environment and positive contribution to human well being.

It would also mean an improved business product, an economical benefit to the consumer (increased miles per gallon) which would be good for business, building a strong brand. It should be noted that fuel injection technology was borne out of the efforts to improve the catalytic converter performance. This technology caught on, making broader use of electronic technology, and has virtually illuminated the use of the carburetor.

A sustainability approach may also be applied in ongoing research and development (R&D) in all industries subject to Tier 3, mainly the petroleum industry, the motor vehicle manufacturing industry and the emissions control equipment industry. And all these can benefit from applying information technology to produce improved products with optimal efficiency.

References

Gerard, David; Lave, Lester b.; Implementing Technology Forcing *Policies; Technology Forecasting and Social Change*, 72 (2005)

www.Jdpowers.com

www.EPA.gov

www.fueleconomy.gov

www.UNEP.org

401748 FCC Gasoline Hydrodesulfurization and U.L.S.G

Abstract:

The USEPA Tier 3 regulation becomes effective in January, 2017. The most attractive options for satisfying the new regulation will involve some combination of the following requirements: least additional octane loss, least capital investment, fastest implementation time, least additional plot space requirement, sufficient availability. This discussion will solely consider selective hydrodesulfurization technology and how it can be utilized to satisfy the new regulation. As will hopefully be made clear in the discussion, it is not often a trivial matter for selective hydrodesulfurization technology designed for the present regulation to be operated so that the new regulation can be satisfied. If minimum additional mono-olefin loss and minimum additional hydrogen consumption have a high priority, then the mercaptan recombination reaction must receive very careful consideration.

REFINING SOLUTIONS IN A CHANGING WORLD

RFG, RFS, SULFUR, BENZENE, TIER 3 AND BEYOND

Thomas R. Hogan, P.E.
Senior Vice President
April 28, 2015

AIChE
American Institute of Chemical Engineers

Turner, Mason & Company
CONSULTING ENGINEERS

Turner, Mason & Company

Privately held – established 1971

Dallas, Texas headquarters

Provides engineering and management consulting services to clients in the upstream, midstream and downstream segments of the oil industry

Areas of expertise relevant to this topic include fuels regulations and supply/demand/pricing forecast in semi-annual Crude and Refined Product Outlook

Presentation Outline

- Introduction
- History of Regulatory Change
- Reformulated Gasoline Impact on Refining
- Feedstock Changes
- Potential Future Changes

Introduction

- Changing fuels specifications due to regulations are large drivers in new/obsolete refinery equipment
- Changing crude availability has caused added investment
- Potential new initiatives, such as shifts to higher octane gasoline, will likely require additional investment
- Refiners ability to respond to additional regulations becomes more limited as the restrictions interact

History of Regulatory Change

- Evolving U.S. Fuels Regulations
- Reformulated Gasoline Program
- Renewable Fuel Program

Evolving U.S. Fuels Regulations

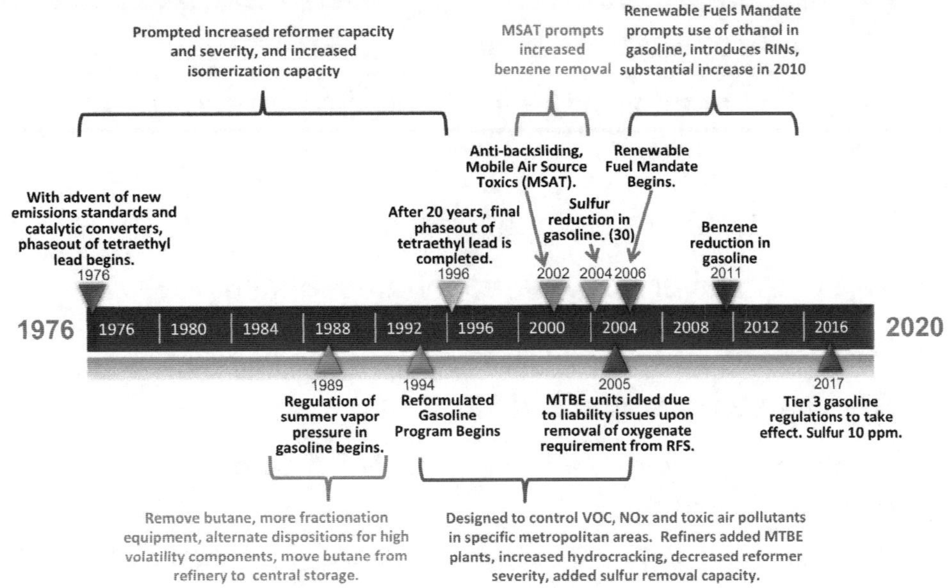

Reformulated Gasoline Program

Revolutionary Program

- Regulatory Negotiation (Reg/Neg), Autos, Oils & EPA
- Complex model – Refinery specific flexibility
- EPA involvement in day to day operations

RFG Program Components

- Establish a baseline
  - ✓ Reviewed by EPA and third parties like TM&C
- Per gallon and/or annual average compliance limits
- Reporting
- Third party laboratory confirmation
- Product transfer documentation and labeling
- Attestations
- RFG survey association
- Contracts between refiners and midstream transportation
- Most components incorporated in subsequent programs

Renewable Fuel in the RFG Program

- Renewable fuel in the RFG program
  - ✓ Mid-continent refiners used ethanol
  - ✓ Most refiners added MTBE units (not a renewable fuel)
  - ✓ MTBE units idled in 2005 due to liability issues when oxygenate requirement was removed from RFG
  - ✓ Essentially all refiners replaced MTBE with ethanol
  - ✓ Lost dilution impact, less ethanol to get same oxygen content (MTBE ~ 18 wt%, Fuel Ethanol ~ 33 wt%)
  - ✓ No significant re-use of MTBE plants
  - ✓ Value of MTBE capacity was ~ $2-3 billion (EIA 3/00)

AIChE 9

Renewable Fuel Program

- Began in 2006/2007
- Mandated 36 billion gallons by 2022
- Most investment was made downstream of the refinery to allow the addition of ethanol at distribution terminals
- Additional octane as ethanol increased to 10 vol%
- Program is in flux

AIChE 10

Reformulated Gasoline Impact on Refining

- Summary of Refining Processes
- Gasoline Properties

AIChE 11

Summary of Refinery Processes

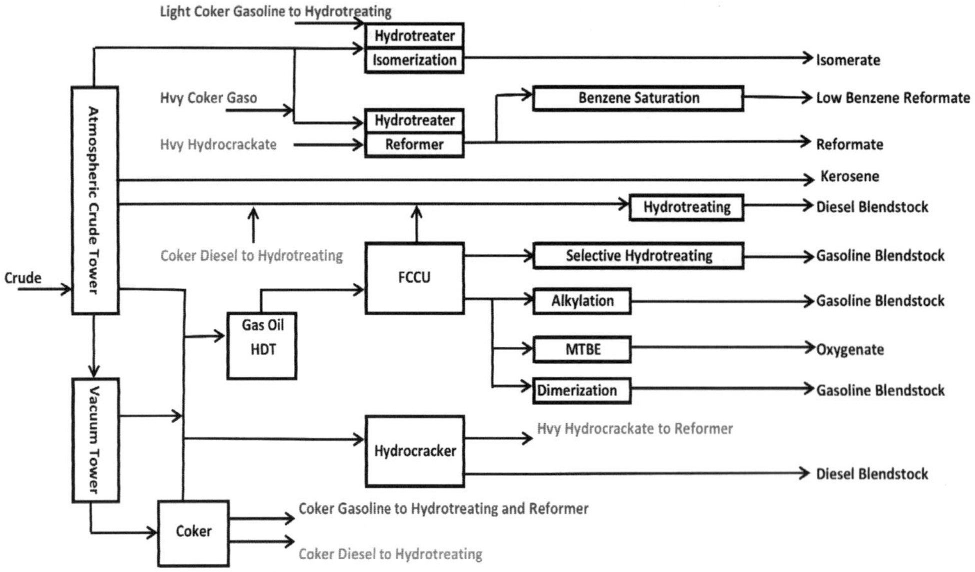

AIChE 12

Gasoline Properties

		Average			Comments
		1990 Baseline		2017	
Property		Summer	Winter	Annual	
RVP	psi	8.7	11.5		~7 for summer RFG
Oxygen	wt%	0	0	~3.5	E15 would be 5.2 wt%
Sulfur	ppm	339	338	10	97% reduction
Benzene	vol%	1.53	1.64	0.62	~60% reduction
Aromatics	vol%	32	26.4		
Olefins	vol%	9.2	11.9		
E200	vol%	41.0	50.0		
E300	vol%	83.0	83.0		

Gasoline Blendstock Properties

Typical 1990								
					FCC		HDC	
Property		LSR	Isom	Reform	Light	Hvy	Lt	Alky
RVP	psi	12-14	14-16	2.5-3.5	8-9	1-2	14	3-5
Sulfur	ppm	50-150	0	0	<900	<4,000	30	10
Benzene	vol%	1.0	0	2-5	1-3	0	~2	0
Aromatics	vol%	1.0	0	55-65	5-8	50-65	~2	0
Olefins	vol%	0	0	0	40-50	10-20	0	0
% of Pool*		5	7	33	18	19	4	8

*Balance of pool is butane ~ 6%

1. LSR, Isomerate, Lt HDC primary sources of RVP after butane
2. FCC gasoline requires treatment for all of EPA properties
3. Reformate and FCC gasolines contribute benzene and aromatics
4. Reformate and FCC gasoline focus of refinery solutions

Refinery Changes for RFG

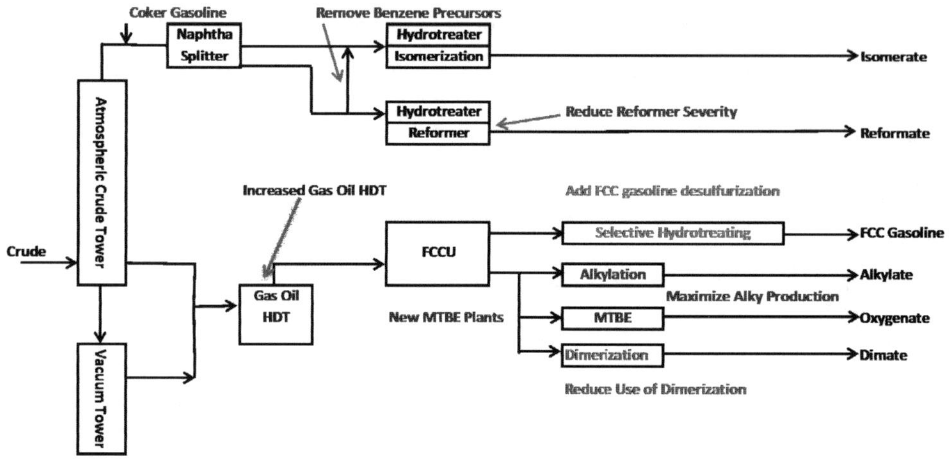

AIChE 15

Changing Crude Slates

- Heavier Crude
- Lighter Crude

AIChE 16

Heavier Crude Slates

- Up until about ten years ago, everyone "knew" that heavier crude oil refiners would be advantaged over those only running light, sweet crude oil
- Most new production at that time had more vacuum bottoms and more sulfur than the typical WTI
- Refiners invested billions of dollars in cokers, FCC pre-treaters, hydrocrackers and de-sulfurizing technology just to produce products to meet pre-RFG specifications
- Heavier crude slates unloaded the parts of the refinery that handled the light and heavy naphtha

AIChE 17

Lighter Crude Slates

- And then came the light, tight oil revolution
- Lighter crude slate refiners became advantaged, especially those close to "stranded" crude
- The light/heavy price differential compressed, and ratio of the LLS price vs. Maya price decreased below 1.2

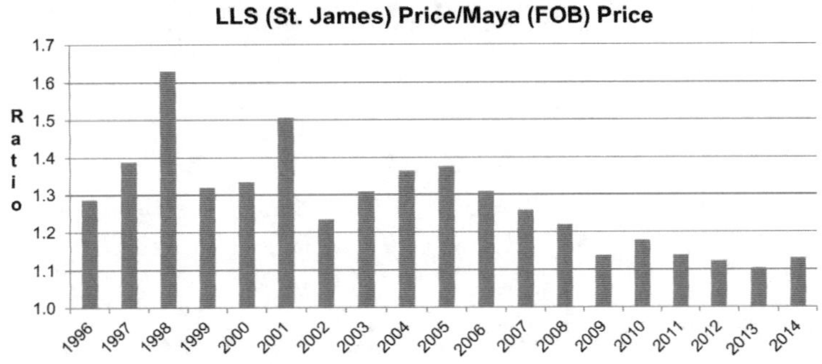

AIChE 18

Lighter Crude Slates (cont)

- Refiners are responding to the lighter crude slates by:
  - ✓ Crude unit modifications
  - ✓ Additional gas compression and gas plant modifications
  - ✓ Additional naphtha splitting and hydrotreating
  - ✓ Additional isomerization capacity
  - ✓ Purchasing feedstocks or new crude capacity to fill FCC, Coker and Hydrocracking units
  - ✓ Light/heavy crude blends to replace medium crudes

Future Changes

- Potential increase in octane

- Impact of interacting regulations

Potential Increase in Gasoline Octane

- Increasing CAFE standards may need to be met by turbocharged engines that require higher octane gasoline
  - ✓ Refining solutions include
    - Increasing reformer severity which has been loafing
    - Increasing isomerization of light straight run gasoline
    - Increasing dimerization
    - Increasing gas liquids conversion of butane/isobutane to isooctane/isooctene
    - Addition of more ethanol/other oxygenate

Limitations on Combination of Regulations

- Reducing aromatics plus higher octane is an example of potential regulatory conflict
  - ✓ Reducing aromatics reduces octane
  - ✓ Although the refiners might be able to accomplish both, it would further limit refining flexibility
  - ✓ It's likely that to meet both limits, all of the refinery blendstocks must be available
  - ✓ Refinery turnarounds will need to be whole refinery shutdowns, which stresses the U.S. refinery system

Conclusions

- Refiners have successfully responded to significant (costly) change over the last 25 years
- From 1990 through 2012, the industry invested about $128 billion[1] to meet regulatory requirements, over 80% of capital spent by refining in that period.
- The number and complexity of regulatory changes has increased dramatically in the past 20 years and have been primarily technology forcing
- Regulations have caused refinery operations to become increasingly complex and must not be evaluated in isolation

1. From July 16, 2013 testimony to Congress by William R. Klesse

Conclusions (cont.)

- Refiners are investing for the future including
  - ✓ Meeting Tier 3 and CAFE standards,
  - ✓ Processing lighter crudes (Est. $4 billion 2014-2020)[1], and
  - ✓ Supplying export markets

1. Oil and Gas Financial Journal, July 22, 2014

FCC Gasoline Desulfurization and Tier 3 Solutions

Arvids Judzis Jr, Director Technology
Stephen Williams, Business Development Manager

CB&I's Lummus Refining and Gasification business unit
The Woodlands, Texas

15 AIChE
Spring Meeting, Austin, TX

CBI

Safe Harbor Statement

This presentation contains forward-looking statements regarding CB&I and represents our expectations and beliefs concerning future events. These forward-looking statements are intended to be covered by the safe harbor for forward-looking statements provided by the Private Securities Litigation Reform Act of 1995. Forward-looking statements involve known and unknown risks and uncertainties. When considering any statements that are predictive in nature, depend upon or refer to future events or conditions, or use or contain words, terms, phrases, or expressions such as "achieve", "forecast", "plan", "propose", "strategy", "envision", "hope", "will", "continue", "potential", "expect", "believe", "anticipate", "project", "estimate", "predict", "intend", "should", "could", "may", "might", or similar forward-looking statements, we refer you to the cautionary statements concerning risk factors and "Forward-Looking Statements" described under "Risk Factors" in Item 1A of our Annual Report filed on Form 10-K filed with the SEC for the year ended December 31, 2014, and any updates to those risk factors or "Forward-Looking Statements" included in our subsequent Quarterly Reports on Form 10-Q filed with the SEC, which cautionary statements are incorporated herein by reference.

15 AIChE
Spring Meeting, Austin, TX

CBI

Agenda

- Marketplace Trends
- Characteristics of FCC Gasoline
- Concept of Selective Desulfurization
- Revamp Approaches

15 AIChE
Spring Meeting, Austin, TX

CBI

Marketplace Trends

- **For Tier 3 compliance, we see ...**
  - Most refiners did not pre–invest in 10 ppm sulfur gasoline, so there will be a need to spend some capital
  - Sulfur credits are being used by many operators to delay capital expenditure
  - Modest capital is being, and will be, spent for revamping existing post treatment units (as late as possible, however)
  - Desulfurization capacity is impacted when co-processing 'excess' straight run naphtha with FCC gasoline
- **Other considerations**
  - Increased tight oil feed can make 10 ppm gasoline targets easier to achieve
  - Alkylate demand increasing due to low sulfur and excellent octane; puts more stress on existing Alkylation facilities

15 AIChE
Spring Meeting, Austin, TX

CBI

Marketplace Trends – Octane Value

- Octane value has increased 50 + % since 2010
- Octane value expected to continue above historical value as a result of Tier 3
- Octane is valued by car manufacturers to meet CAFE standards
- Octane retention during HDS treatment now more important than ever

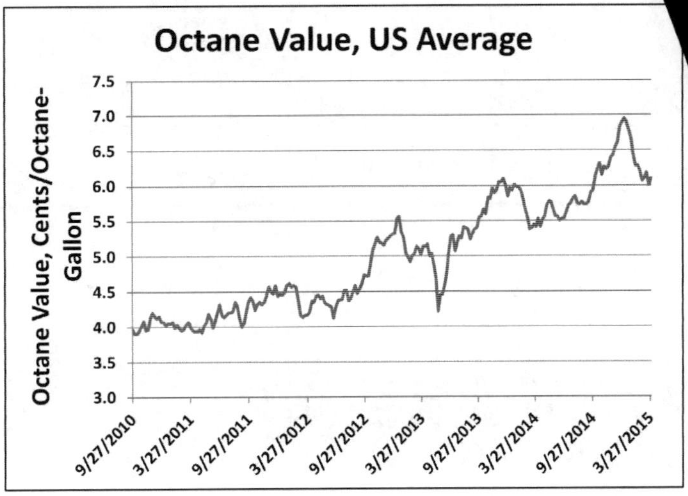

US Energy Information Administration – data through 3/30/15

15 AIChE Spring Meeting, Austin, TX

CBI

FCC Gasoline Desulfurization

Most Important Factors:

- Process Safety
- Minimal Octane Loss
- Minimal Hydrogen Consumption
- Flexibility
  - Long Catalyst Life (meet/ exceed FCC T/A cycle)
  - Feedstock/Product Flexibility
  - High Reliability
- Low Investment Costs

15 AIChE Spring Meeting, Austin, TX

CBI

Typical FCC Gasoline Properties

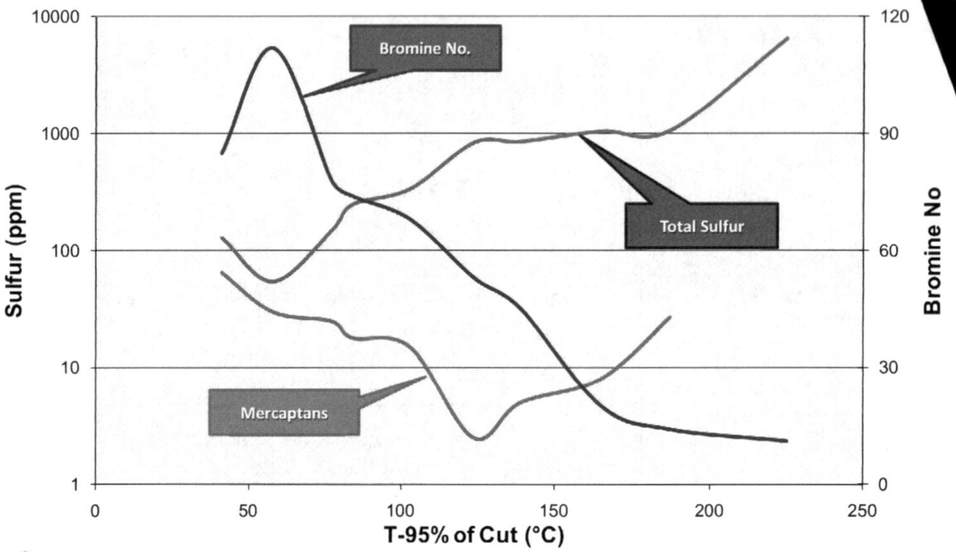

15 AIChE
Spring Meeting, Austin, TX

CBI

Treatment of Light Cat Naphtha

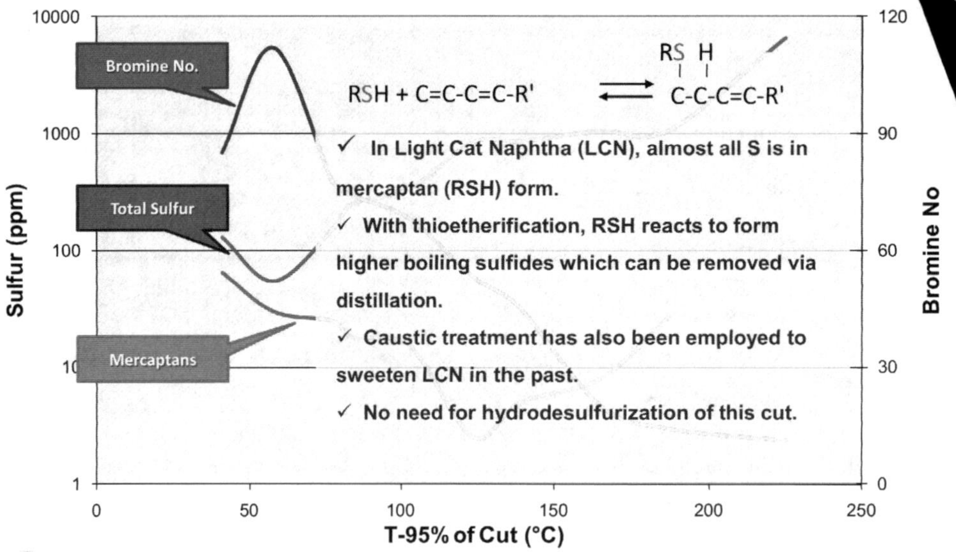

RSH + C=C-C=C-R' ⇄ C(SR)-C(H)-C=C-R'

- ✓ In Light Cat Naphtha (LCN), almost all S is in mercaptan (RSH) form.
- ✓ With thioetherification, RSH reacts to form higher boiling sulfides which can be removed via distillation.
- ✓ Caustic treatment has also been employed to sweeten LCN in the past.
- ✓ No need for hydrodesulfurization of this cut.

15 AIChE
Spring Meeting, Austin, TX

CBI

Hydrodesulfurization of HCN

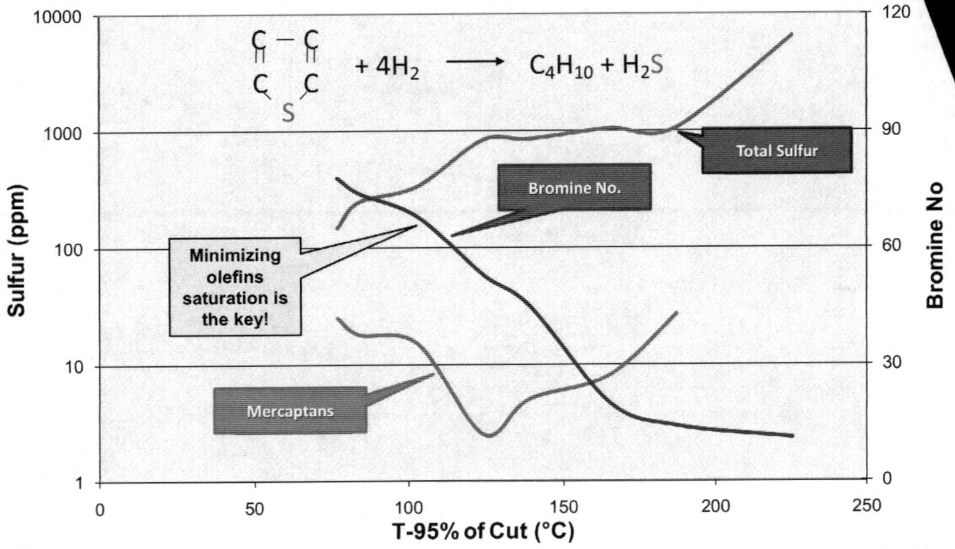

15 AIChE Spring Meeting, Austin, TX

CBI

Desulfurization Strategy - CD*Hydro*® /CD*HDS*®

Treatment of each fraction with minimum possible severity:

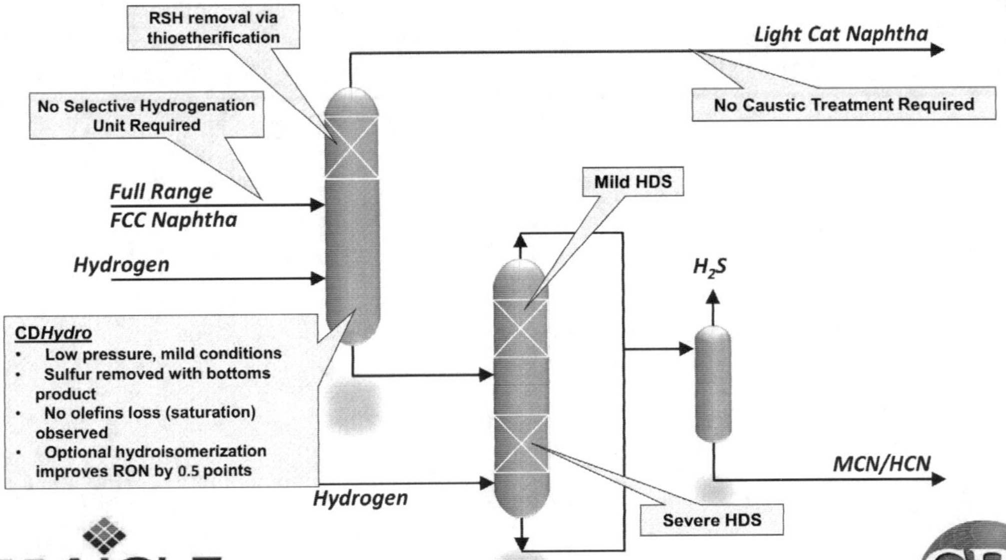

15 AIChE Spring Meeting, Austin, TX

CBI

CDHDS Selectivity/ Octane Retention

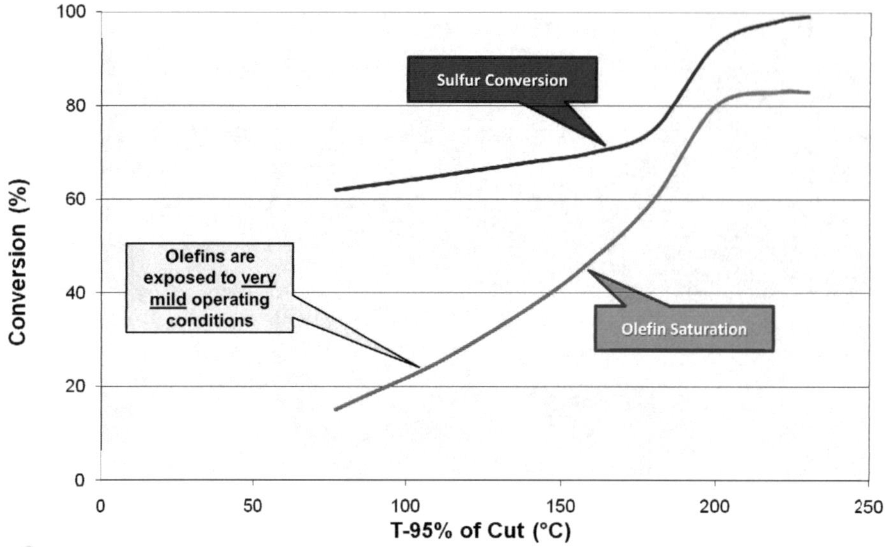

15 AIChE Spring Meeting, Austin, TX

CB&I

Conventional Fixed Bed Reactor

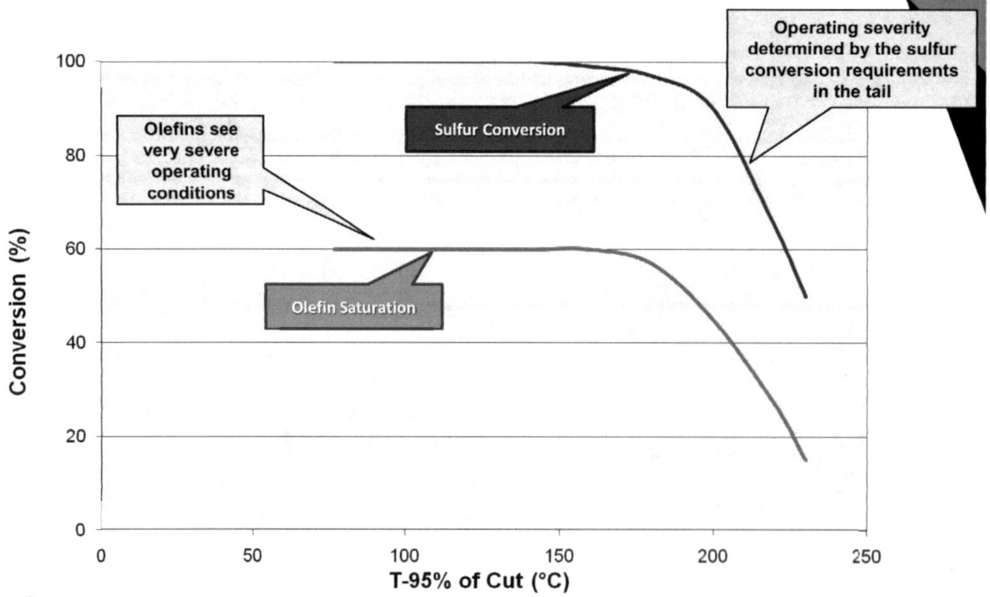

15 AIChE Spring Meeting, Austin, TX

CB&I

Tier 3 Desulfurization Strategy

with CD*Hydro*/ CD*HDS* Technology

Typical Solutions:

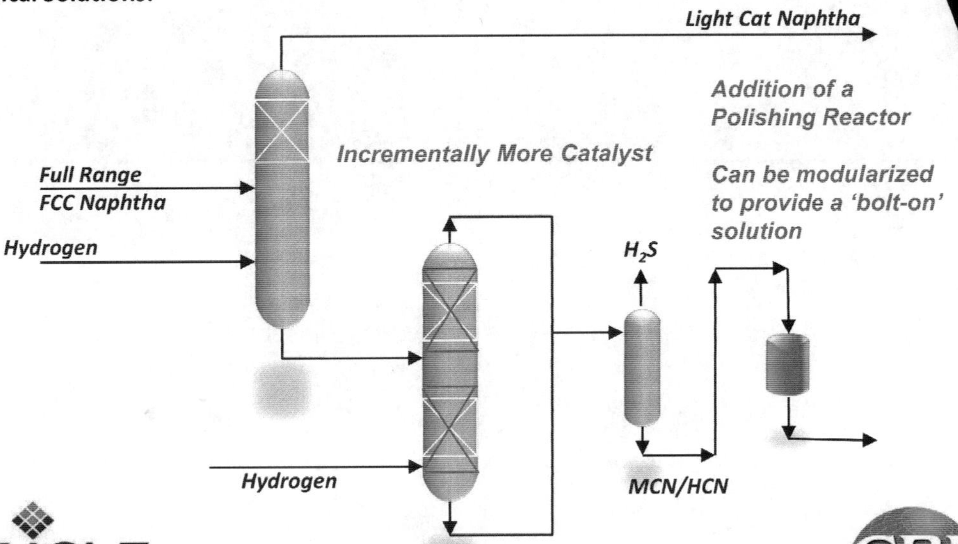

15 AIChE Spring Meeting, Austin, TX

CB&I

Tier 3 Desulfurization Strategy

with an Existing Fixed-Bed HDS Unit

Add a CD*Hydro* Column:

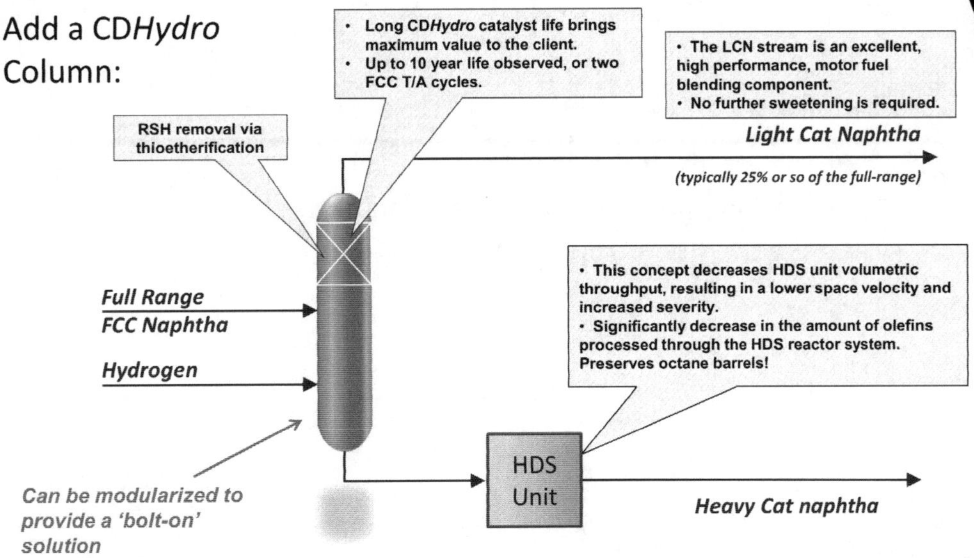

Existing 'Full-Range' Naphtha Hydrodesulfurization Unit

15 AIChE Spring Meeting, Austin, TX

CB&I

Modularized Products

CB&I's Engineered Products in Tyler, Texas

15 AIChE Spring Meeting, Austin, TX

CB&I

References for FCC Gasoline CD*Hydro*/CD*HDS*

- More than 120 licenses
- Nearly 1.9 MM Bbl/day capacity
- Seven units have demonstrated ability to produce cat gasoline <10 ppm sulfur

15 AIChE Spring Meeting, Austin, TX

CB&I

Questions and Answers

CONTACTS	
Arvids Judzis, Jr. Director Technology CB&I's Lummus Refining and Gasification The Woodlands, Texas Office: +1 832 513 1388 Email: ajudzis@cbi.com	Stephen D. Williams, P.E. Technology Business Development Manager CB&I's Lummus Refining and Gasification The Woodlands, Texas Office: +1 832 513 1502 Email: sdwilliams@cbi.com

15 AIChE
Spring Meeting, Austin, TX

15 AIChE
Spring Meeting, Austin, TX

Tier 3 Processing Options for Natural Gasoline from NGL Fractionation

Presentation ID 400173
April 28, 2015
by

Dominic Varraveto, PE
Chief Process Engineer
&
Director of Process Technology

15 AIChE
Spring Meeting, Austin, TX

BURNS McDONNELL

1. Introduction & Agenda
2. NGL Fractionation
3. Natural Gasoline
4. Tier 3 Regulation
5. Natural Gasoline
6. Processing Options
7. Summary
8. Q & A

15 AIChE
Spring Meeting, Austin, TX

BURNS McDONNELL

2. NGL Fractionation

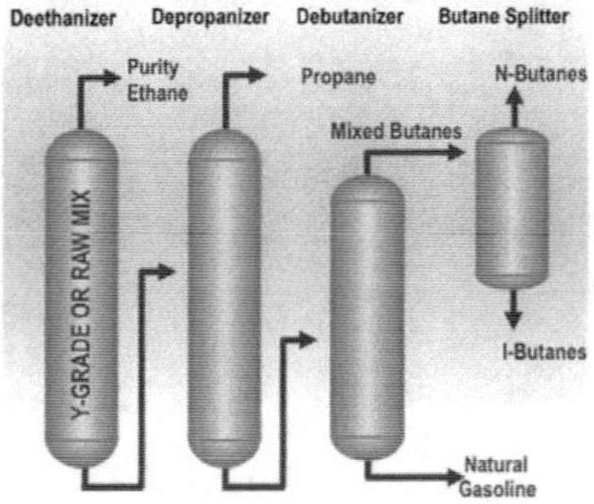

UOP

15 AIChE Spring Meeting, Austin, TX

BURNS & McDONNELL

2.1 NGL Fractionation Process

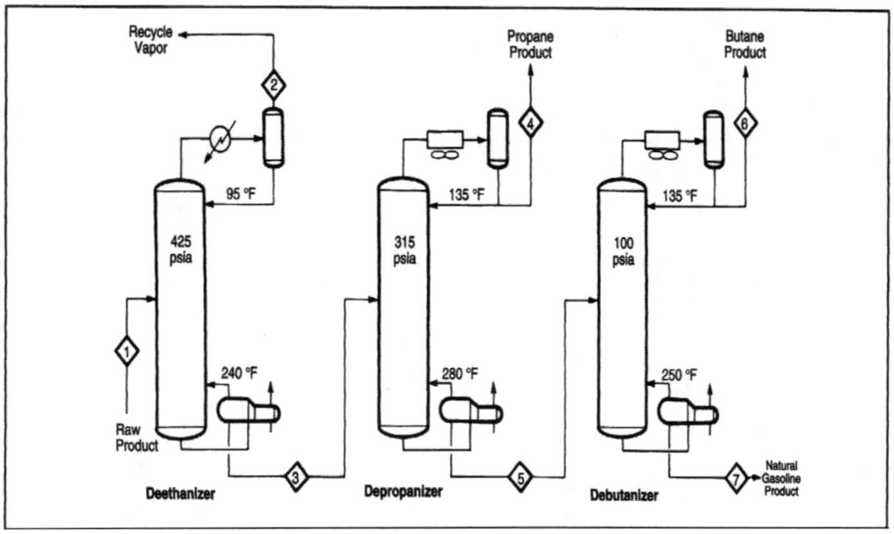

15 AIChE Spring Meeting, Austin, TX

BURNS & McDONNELL

2.2 NGL Fractionation Feed

Y Grade

- De-methanized, liquid feed containing
- Ethane, Propane, Butane
- Pentane + (aka Natural Gasoline)
- Limited: Methane Olefins, Aromatics
- Contaminants: Sulfur, Carbon Dioxide, Water
- Properties: TVP, Distillation EP, Corrosiveness

Trends

- Increasing volumes
- De-ethanized
- Higher fraction of heavier components

15 AIChE
Spring Meeting, Austin, TX

BURNS McDONNELL

2.3 US NGL Fractionation

- No of plants
- Range of throughput
- Typical plant size
  - 75k BPD
- Estimated NGL Fractionation Industry
  - 2,000k BPD
- Pentane +
  - 10-15%
  - 200 -300k BPD

15 AIChE
Spring Meeting, Austin, TX

BURNS McDONNELL

3. Natural Gasoline

Natural Gasoline cannot be used to fill your tank until it is processed and combined with other gasoline blend stocks.

15 AIChE Spring Meeting, Austin, TX

BURNS McDONNELL

3.1 Natural vs Finished Gasoline

Natural Gasoline

- Low Octane
- High Vapor Pressure
- Lacks Approved Oxygenate
- 100-200 ppm Sulfur
- Low olefins
- Paraffinic
- Low Benzene & Aromatics
- Low T50
- Low T90

Finished gasoline

- Octane: (R+M) / 2
- Vapor Pressure: RVP
- Oxygenate
- Sulfur
- Olefins
- Benzene
- Aromatics
- T50
- T90

15 AIChE Spring Meeting, Austin, TX

BURNS McDONNELL

4. Tier 3 Regulations

15 AIChE
Spring Meeting, Austin, TX

BURNS & McDONNELL

4.1 Tier 3 Overview

- Tier 3 is the latest EPA rule designed to further reduce air pollution
- It is applicable to light duty vehicles - passenger cars and trucks, fueled with gasoline.
- Considers fuel and vehicle as an integrated system
- Aligned with CARB LEV III so same vehicles can be sold in all 50 states
- Allows vehicle emission control systems to operate more efficiently

4/15/2015

15 AIChE
Spring Meeting, Austin, TX

BURNS & McDONNELL

4.2 Tier 3 Fuel Regulations

- EPA published final rule
  - April 28, 2014
- Sulfur, ppmw
  - 10 annual average / **80** max refinery gate
  - Retains E10 (10% Ethanol) as test fuel
- Effective Date Jan 1, 2017
  - Implementation time not extended
- **Averaging, Banking & Trading, ABT**
- Concessions for small refiners & **small refineries**

15 AIChE Spring Meeting, Austin, TX

BURNS McDONNELL

4.3 Purpose of Tier 3 Regulation

- Addresses public health issues
  - Requested by May 21, 2010 Presidential Memo
- Significantly reduces light-duty emissions
  - NOX, VOC,CO and Air Toxics (i.e. benzene) by 80% of fleet avg
  - PM 2.5 by 70% per vehicle
- Facilitates development of lower cost technology to improve fuel economy
  - Reduces Green House Gases (GHG) emissions
- Extends required warranty life of emission control systems
  - From 120k to 150k miles

15 AIChE Spring Meeting, Austin, TX

4/15/2015

BURNS McDONNELL

4.4 Projected Benefits in 2030

Per the EPA by 2030 Tier 3 standards will **prevent**

- 770 to 2,000 premature deaths
- 2,200 hospital admissions and asthma related ER visits
- 19,000 asthma exacerbations
- 30,000 upper and lower respiratory symptoms in children
- 1.4 million lost school days, work days and minor restricted activities

$7 billion to $19 billion in annual health benefits

15 AIChE 4/15/2015
Spring Meeting, Austin, TX

BURNS & McDONNELL

5. Natural Gasoline Characteristics

- Composition
- Sulfur
- Vapor Pressure
- Octane

15 AIChE
Spring Meeting, Austin, TX

BURNS & McDONNELL

5.1 Natural Gasoline Composition

Component	B.P., °F	Volume %
iso-Butane	11	0
n-Butane	31	1
iso-Pentane	82	30
n-Pentane	97	34
C6's	120-180	25
C7's	170-230	7
C8+	210-290+	3

15 AIChE Spring Meeting, Austin, TX

BURNS & McDONNELL

5.2 Natural Gasoline Sulfur

Total Sulfur, ppmw S:
136/162 /96/25
(Avg/Max/Min/Std Dev)
Sulfur Speciation, wt%:
Extractable
  Mercaptan, 41
Non-Extractable
  Sulfide, 8
  Di-sulfide, 46
  Other, 5

- Mercaptans, RSH
- Sulfides, H2S, R-S-R
- Disulfides, HRS-SRH
- Other -Thiophenes

15 AIChE Spring Meeting, Austin, TX

BURNS & McDONNELL

5.3 Natural Gasoline RVP & Octane

Component	RVP	Research	Motor
iso-Butane	72.2	102.1	97.0
n-Butane	51.6	94.0	89.1
iso-Pentane	20.4	93.0	89.7
n-Pentane	15.6	61.8	63.2
n-Hexane	4.96	24.8	26
n-Heptane	1.62	0	0
iso-Octane	1.71	100	100
Benzene	3.22	101	-
Toluene	1.03	119.7	109.1
Ethanol	0.9	113	-

15 AIChE
Spring Meeting, Austin, TX

BURNS McDONNELL

6. Processing Options

- Sweetening
- Extraction
- Fractionation
- Hydrotreating

15 AIChE
Spring Meeting, Austin, TX

BURNS McDONNELL

6.1 Sweetening

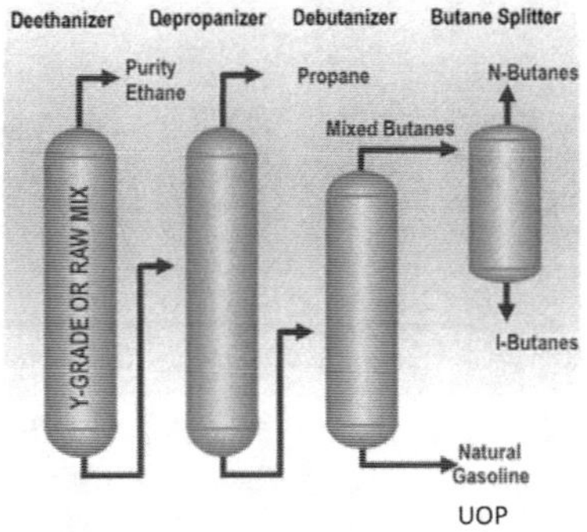

Reactive Sulfur is converted to disulfides but left in the product. This is ineffective for meeting Tier 3 requirements .

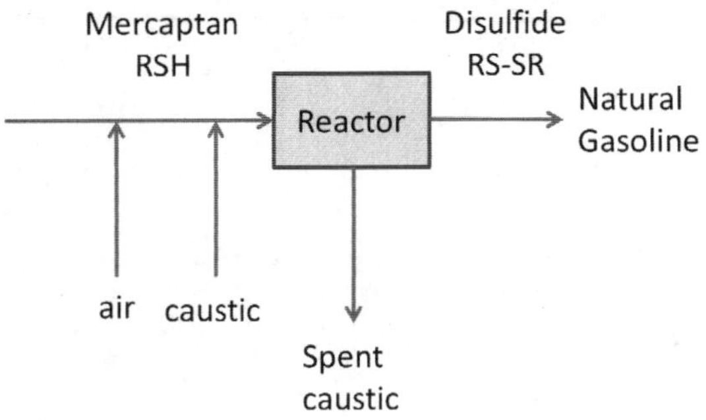

15 AIChE Spring Meeting, Austin, TX

BURNS & McDONNELL

6.2 Extraction

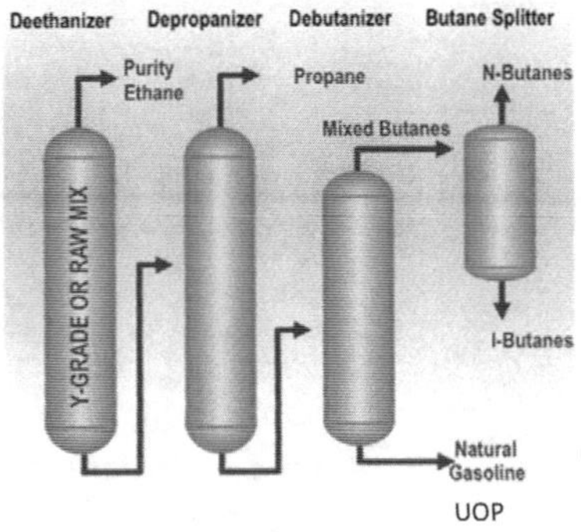

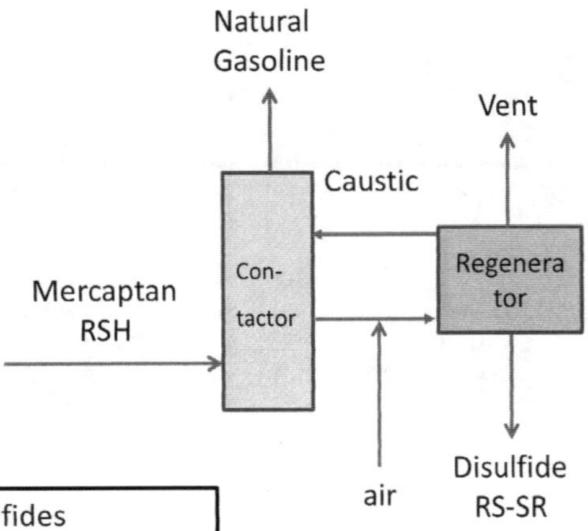

Mercaptan Sulfur is converted to disulfides **and** removed from the product. This is effective for meeting Tier 3 requirements if non-extractable sulfur is not present.

15 AIChE Spring Meeting, Austin, TX

BURNS & McDONNELL

6.3 Fractionation

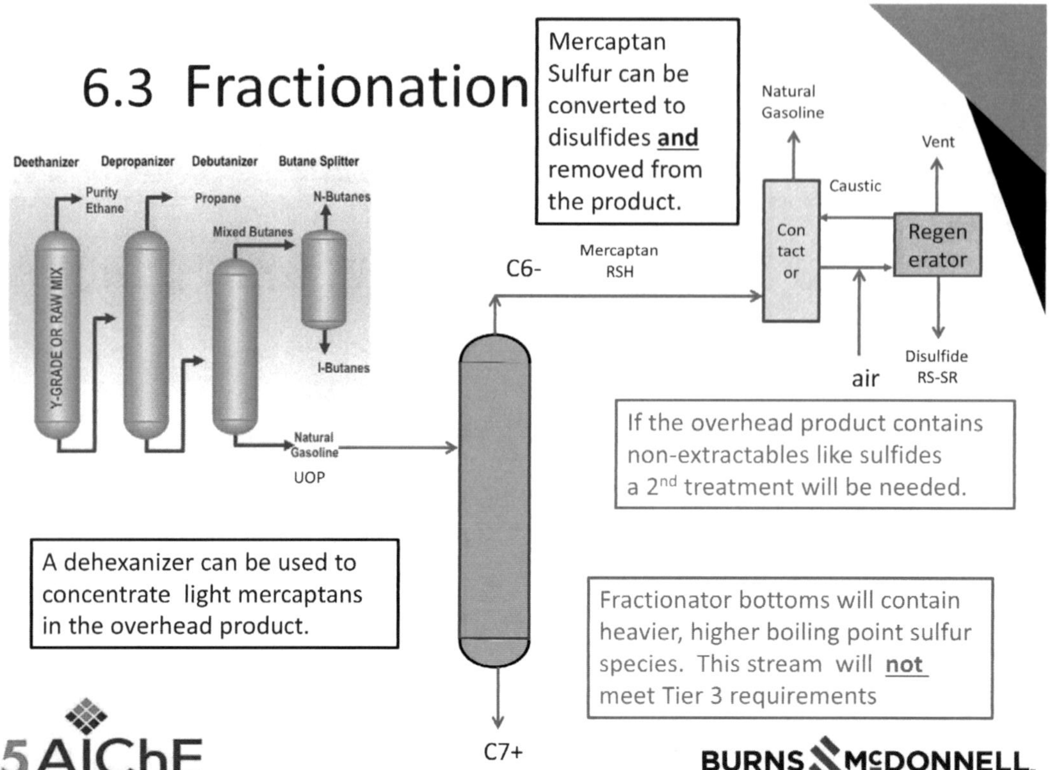

15 AIChE Spring Meeting, Austin, TX

BURNS McDONNELL

6.4 Hydrotreating

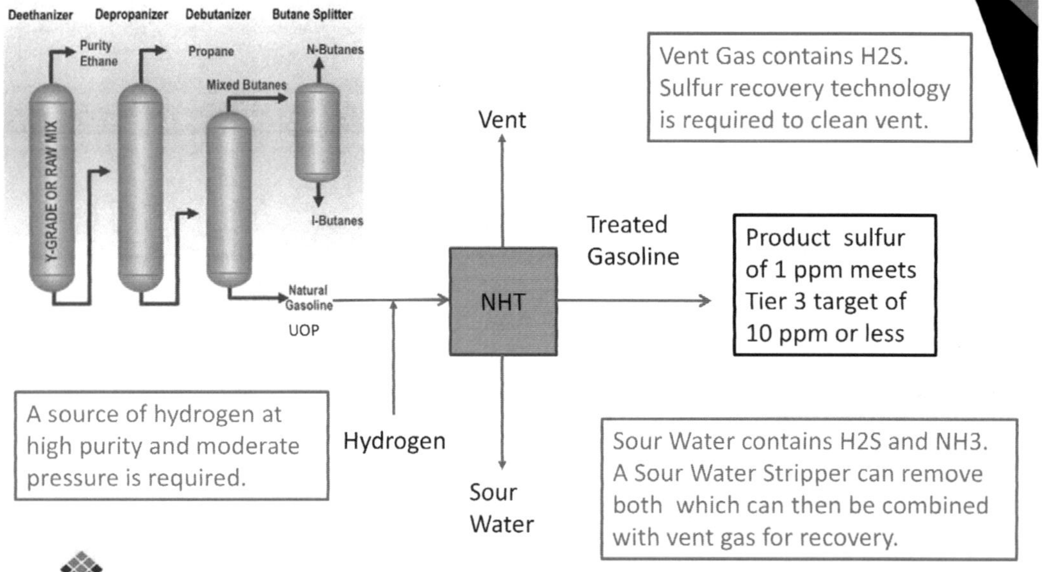

15 AIChE Spring Meeting, Austin, TX

BURNS McDONNELL

7. Summary

- Tier 3 requires sulfur removal from Natural Gasoline to make a sulfur neutral blendstock.
- Traditional Sweetening does not remove Sulfur and is ineffective for meeting Tier 3
- Extraction using caustic removes light mercaptans but is ineffective for removing other sulfur species.
- Fractionating Natural Gasoline to separate extractable from non-extractable sulfur depends on distribution of sulfur species.
- Hydrotreating is effective at removing sulfur to meet Tier 3 but requires additional infrastructure.

15 AIChE Spring Meeting, Austin, TX

BURNS & McDONNELL

8. Q & A

15 AIChE Spring Meeting, Austin, TX

BURNS & McDONNELL

Thank You for Attending

- Burns & McDonnell – 100% Employee Owned!

Dominic Varraveto
dvarraveto@burnsmcd.com
816.822.4282

15 AIChE
Spring Meeting, Austin, TX

BURNS & McDONNELL

Maintaining Flexibility in a Tier 3 Environment

Geoffrey Dubin
Technology Manager
Axens North America

15 AIChE Spring Meeting, Austin, TX

Axens IFP Group Technologies

Agenda

- Tier 3 Overview
- FCC Feed Pre-treatment Challenge
- FCC Post-Treatment Solutions
- Leveraging Tier 3 for Additional Flexibility
- Conclusion

15 AIChE Spring Meeting, Austin, TX

Axens IFP Group Technologies

Tier 3 Regulations

EPA released Tier 3 Final Rulemaking on April 28, 2014

- Average gasoline pool S: **10 wppm**
- Per gallon S cap: **80 wppm**
- Implementation by Jan. 1, 2017
- Extension for small refineries (< 75kBPD) until Jan. 1, 2020

15 AIChE Spring Meeting, Austin, TX

Axens IFP Group Technologies

Tier 3 Approach: 10 ppm S in Pool

- 10-25 ppm S limit in FCC naphtha required
- Other naphtha streams with S can no longer blend directly to gasoline pool (e.g. LSR)
- Many factors affect optimum solution
  - Adequacy of existing CFHT & post-treater
  - Future sour crude slate?
  - Octane value
  - H2 availability
  - Solution in sync with FCC cycle
  - Targeted product slate

15 AIChE Spring Meeting, Austin, TX

Axens IFP Group Technologies

FCC Feed Pretreatment & Tier 3: Deep CFHT Can be a Solution

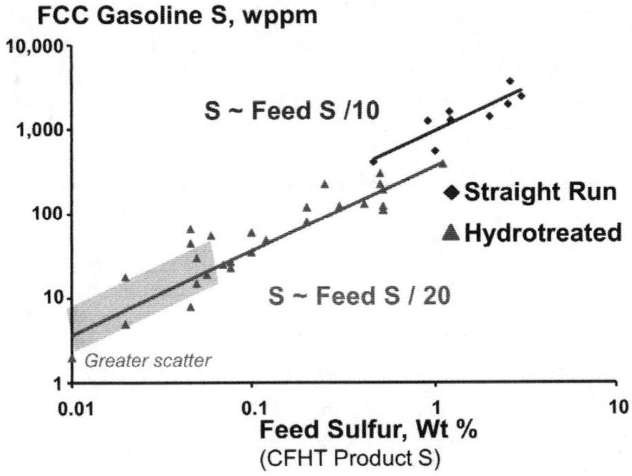

- Lower CFHT Prod S = Shorter CFHT Cycle
- FCC gasoline S v FCC feed S not a simple rule
- CFHT improves FCC feed quality
- Where is the balance?

Less than ~300 ppm S FCC feed required to avoid post-treatment

15 AIChE Spring Meeting, Austin, TX

Axens IFP Group Technologies

CFHT Pretreatment Only Option: Severe Performance for Tier 3 Compliance

CFHT	Existing	New / Modified
Regulation	Tier 2	Tier 3
Pool S Limit, ppm	30	10
FCCN S Limit, ppm	80	20-25
FCC Feed S Limit, ppm	1600	300
CFHT %HDS	91%	98.3%
Cycle Length, month	36	16
Relative Cat Volume for Constant Cycle Length	100%	200%

Example Assumptions:
- Untreated VGO feed = 1.8 w% S, CFHT Pressure = 1200 psig
- S in pool 10% below regulatory limit
- FCCN = 35% of gasoline pool volume
- Standard cut FCC gasoline

15 AIChE Spring Meeting, Austin, TX

Axens IFP Group Technologies

Options for Refineries with CFHT

- Maintain pre-treat alone
  - FCC Feed S < 300 ppm to avoid post-treat
  - Severely reduced cycle length and/or costly revamp
  - No margin for error
    - Cycle length, operation, fractionation (FCCN EBP)
- Revamp CFHT to MHC
  - Deep HDS plus mid distillate production
  - Crude and product flexibility
- Install or revamp post-treater
  - Usually the lowest cost option
  - Co-processing of other streams (LSR, LKN)
  - Pre-treat AND Post-treat for maximum flexibility

15 AIChE Spring Meeting, Austin, TX

Axens IFP Group Technologies

FCC Gasoline Sulfur Profile

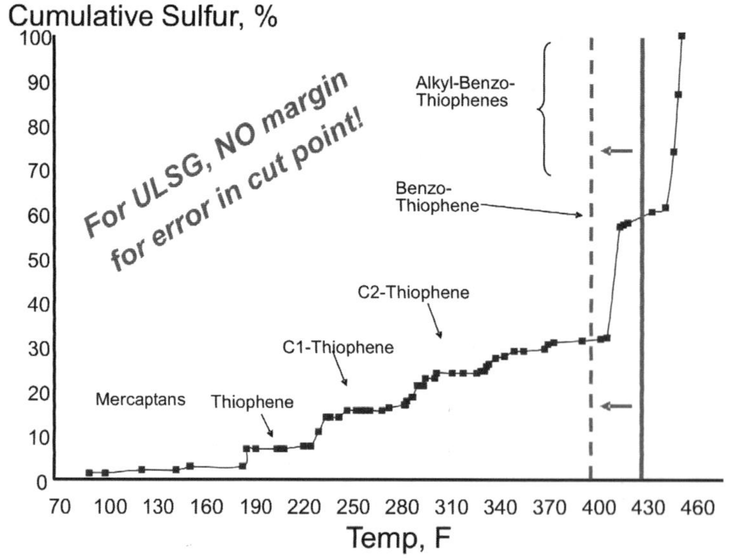

15 AIChE Spring Meeting, Austin, TX

Axens IFP Group Technologies

Focusing in on the FCC Gasoline Composition

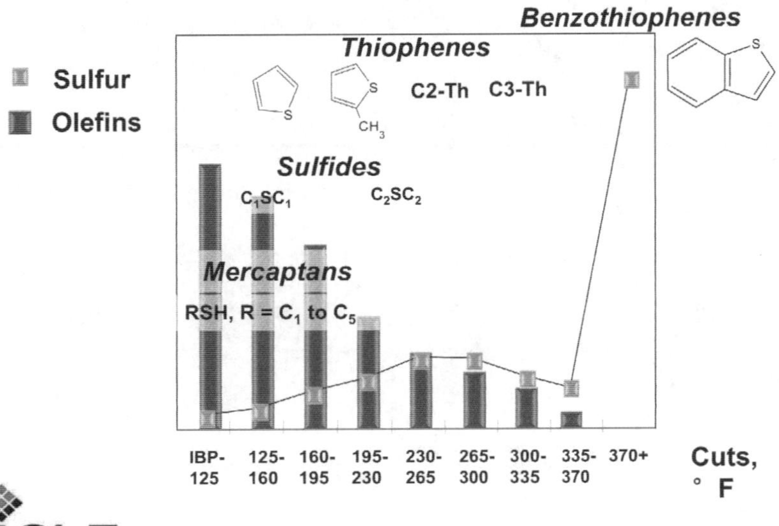

15 AIChE Spring Meeting, Austin, TX

Axens IFP Group Technologies

The Prime-G+™ Technology: Typical Scheme

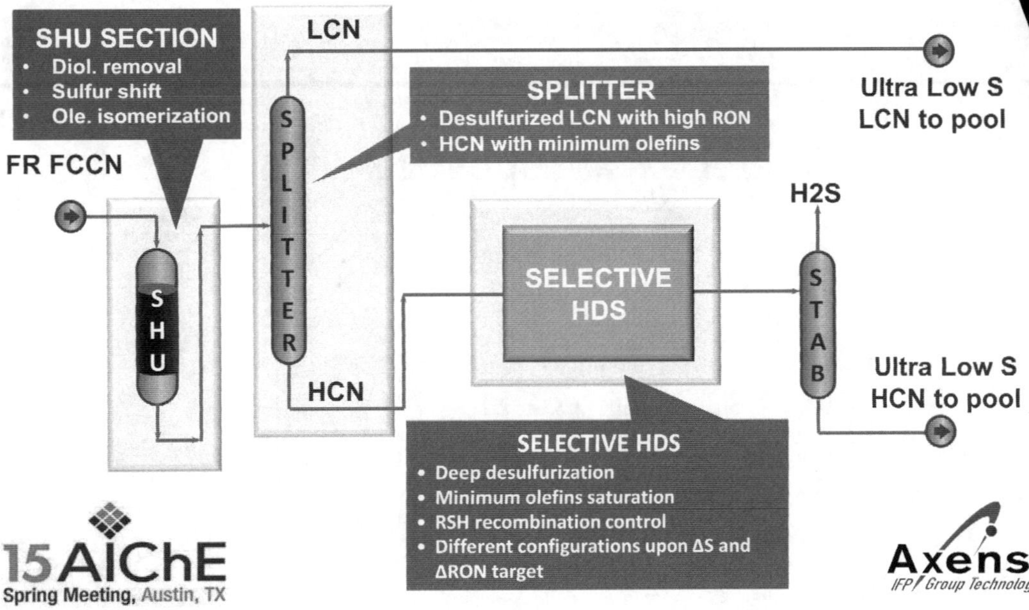

15 AIChE Spring Meeting, Austin, TX

Axens IFP Group Technologies

Tailored Processing Schemes with Prime-G+

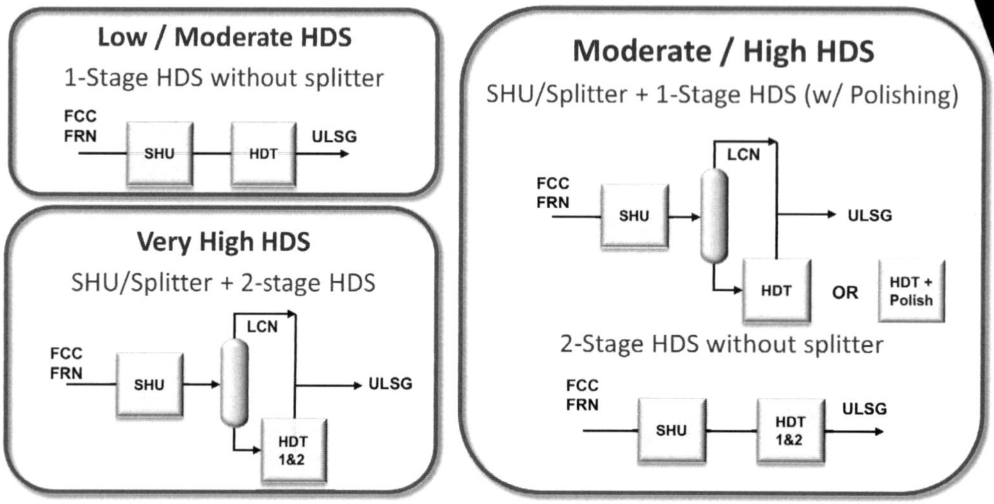

15 AIChE Spring Meeting, Austin, TX

Axens IFP Group Technologies

Prime-G+ Operational Experience

- > 240 References
- > 6MBPSD cumulated capacity
- > 125 References at 10ppm gasoline

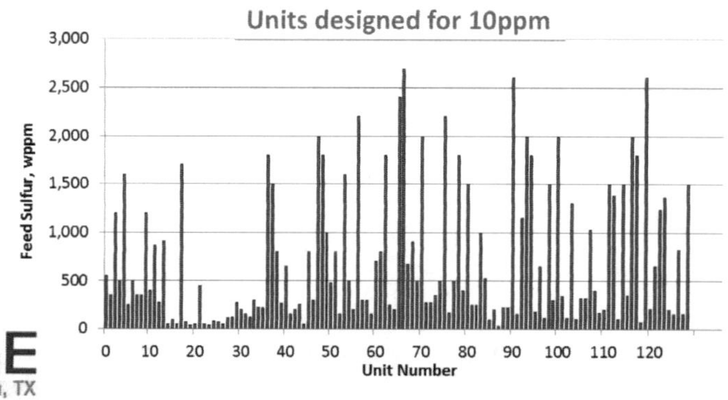

15 AIChE Spring Meeting, Austin, TX

Axens IFP Group Technologies

Tier 3 Post-Treatment Basic Strategy

- Basic strategy for existing post-treaters
  - If splitter, modify LCN cut point to reduce LCN S
  - Increase severity in HDS reactor(s)
- “Do Nothing” consequences
  - Reduced cycle length
  - Increased octane loss
  - Increased Hydrogen consumption
- Multiple Tier 3 solutions exist
  - Feed tailoring (undercut FCCN)
  - Improved catalysts
  - Co-processing of orphan streams
  - Process scheme modification

15 AIChE Spring Meeting, Austin, TX

Axens IFP Group Technologies

Post-Treater Revamp Options (I/III)

- For low-severity post-treaters without a splitter, two main pathways exist;
  - Addition of a splitter
  - Addition of a 2nd HDS stage

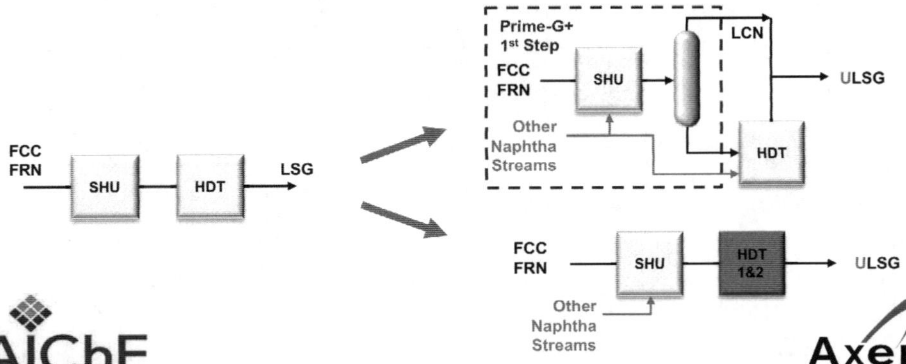

15 AIChE Spring Meeting, Austin, TX

Axens IFP Group Technologies

Post-Treater Revamp Options (II/III)

- For moderate to high severity post-treaters with a splitter, two main pathways exist;
  - Addition of a polishing reactor
  - Addition of a 2nd HDS stage

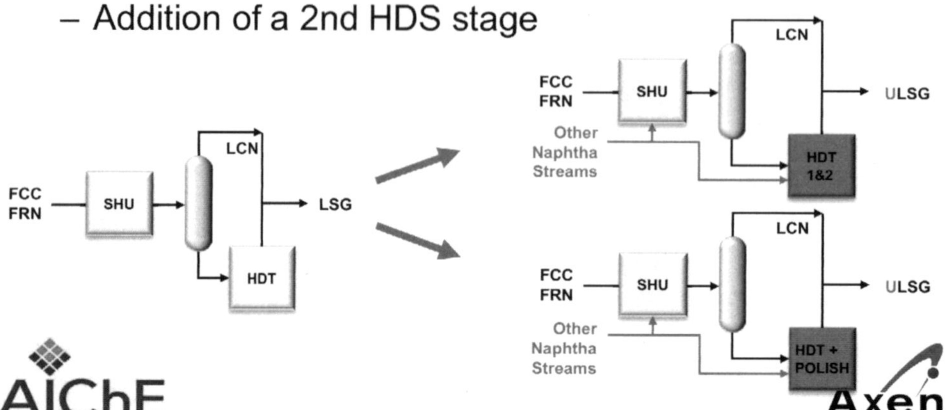

15 AIChE
Spring Meeting, Austin, TX

Axens
IFP Group Technologies

Post-Treater Revamp Options (III/III)

- For post-treaters on the HCN portion of the FCC naphtha
  - Conversion to Prime-G+ 1st Step is attractive solution
  - Revamping sweetening to extractive sweetening for Tier 3 is not required.
  - The Prime-G+ 1st Step allows for the existing caustic sweetening unit to be shutdown

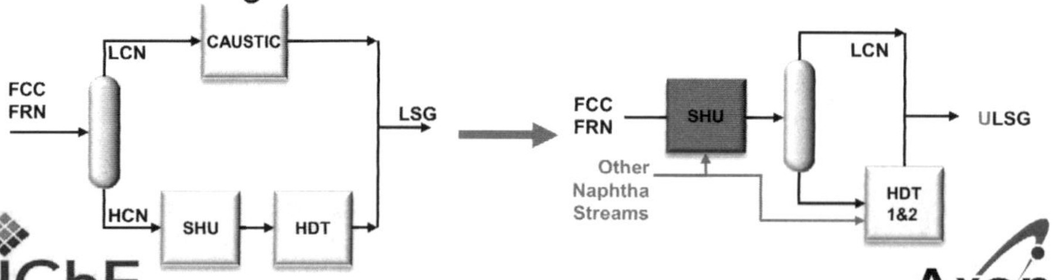

15 AIChE
Spring Meeting, Austin, TX

Axens
IFP Group Technologies

ULSG Despite Varying Feed S

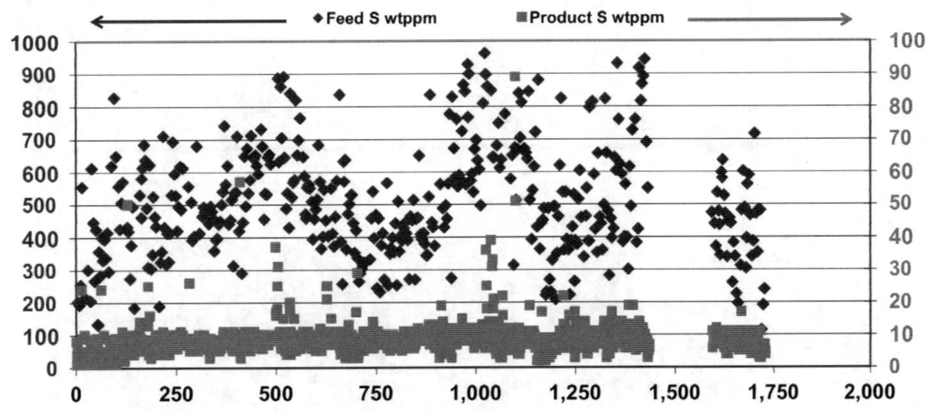

10 ppm operation achieved at 98-99% HDS

15 AIChE Spring Meeting, Austin, TX

Axens IFP Group Technologies

Prime-G+ 1st Step Allows ULS LCN

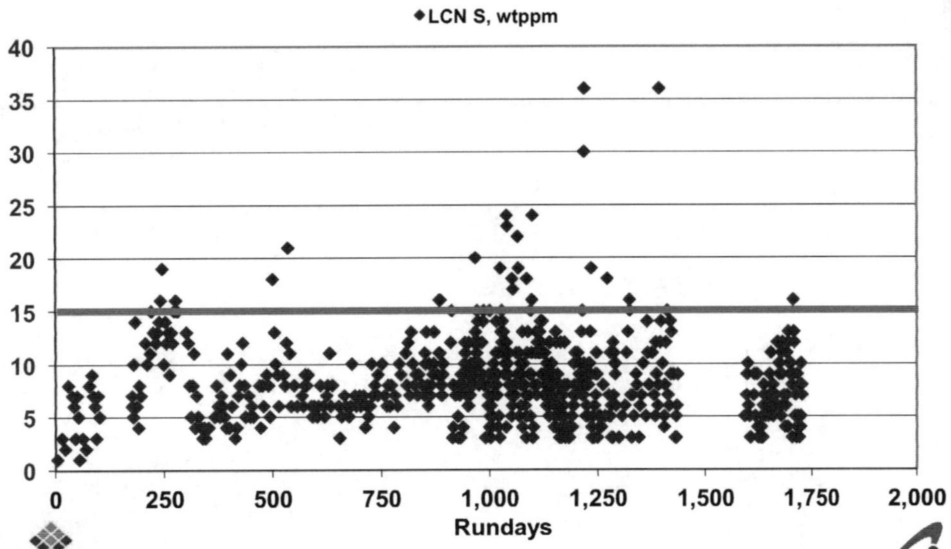

15 AIChE Spring Meeting, Austin, TX

Axens IFP Group Technologies

10ppm Operation at High HDS

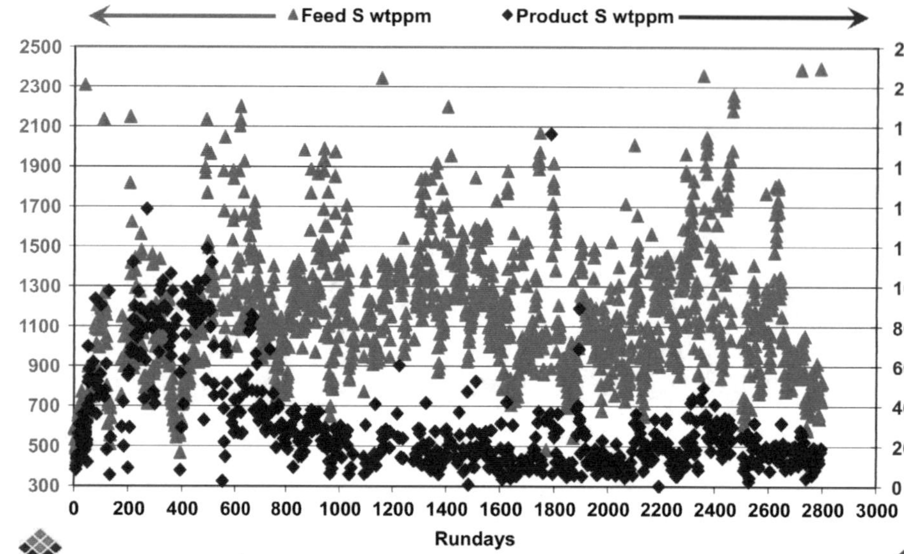

⇨ Stable operation over 7 years

15 AIChE
Spring Meeting, Austin, TX

Axens
IFP Group Technologies

Prime-G+ 1st Step and Addition of 2nd HDS Stage

Phase 1 LCN sulfur average 20 wppm

Phase 2 requiring ULSG production

LCN sulfur average 5 wppm

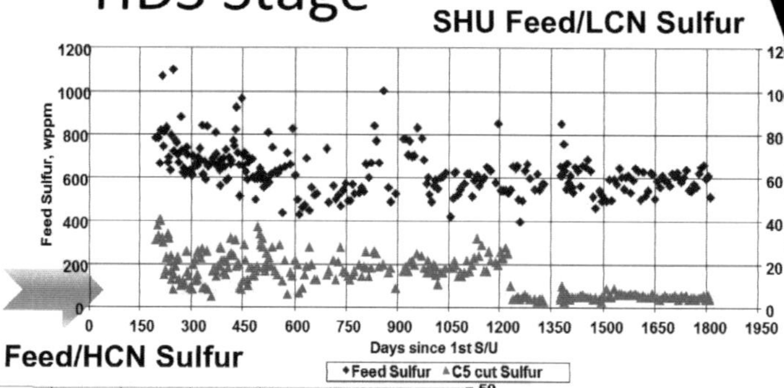

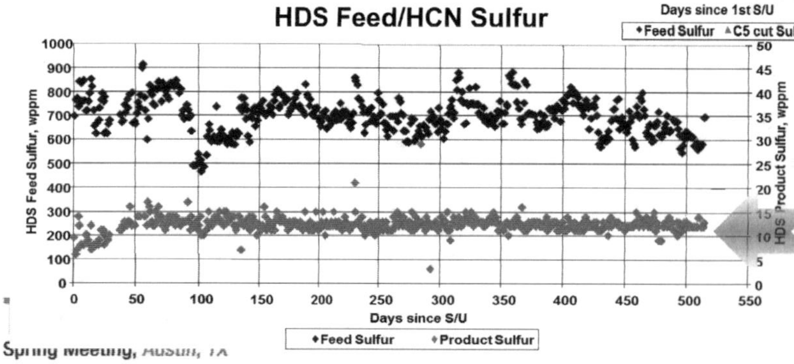

Phase 2 requiring ULSG production

HCN sulfur average 12 wppm

Spring Meeting, Austin, TX

Axens
IFP Group Technologies

Prime-G+ 1st Step with High Feed Sulfur and High RSH

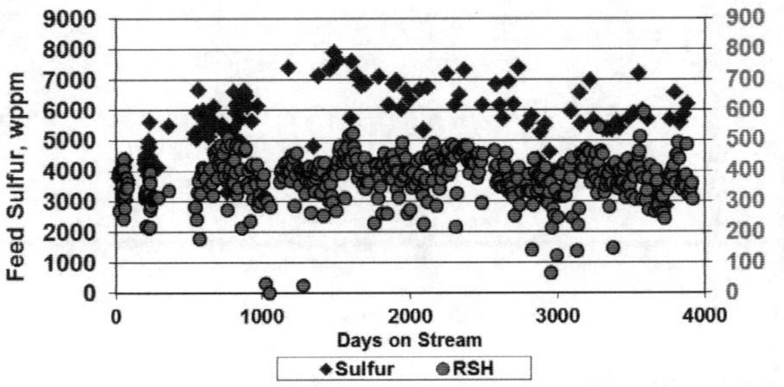

Unit with higher than typical Feed S & RSH content

Feed S ranges from 0.4 – 0.8 wt% and RSH from 300 – 500 wppm

Light RSH conversion after 10 years is 95%+!

LCN RSH content typically near 10 wppm

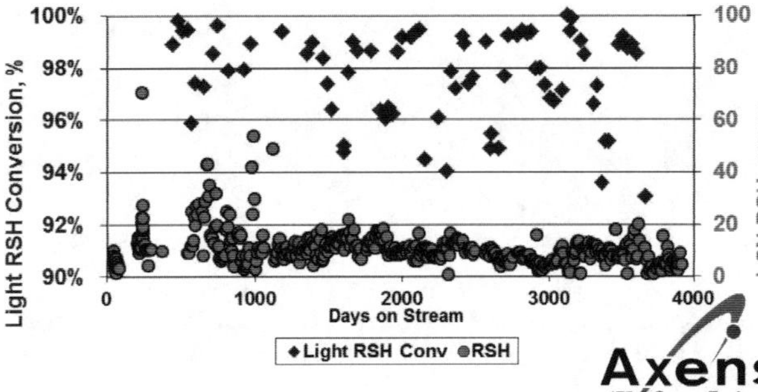

15 AIChE Spring Meeting, Austin, TX

Axens IFP Group Technologies

CFHT Combined with Post-Treating

- Final control of FCC Gasoline S
  - Independent of CFHT operation / upsets
  - When CFHT down, Prime-G+ can operate with higher Feed S

  } Flexibility

  Product S and FCCN end-point control at all times } Reliability
- Post-treater cycles in sync with FCC
- Possibility to co-process other streams
  - SRN, Coker Naphtha, Heavy PyGas
- Ability to optimize CFHT performance/cycle by allowing Prime-G+ to absorb high sulfur swings

15 AIChE Spring Meeting, Austin, TX

Axens IFP Group Technologies

Canadian Heavy Crude Refinery Prime-G+ w/ upstream CFHT

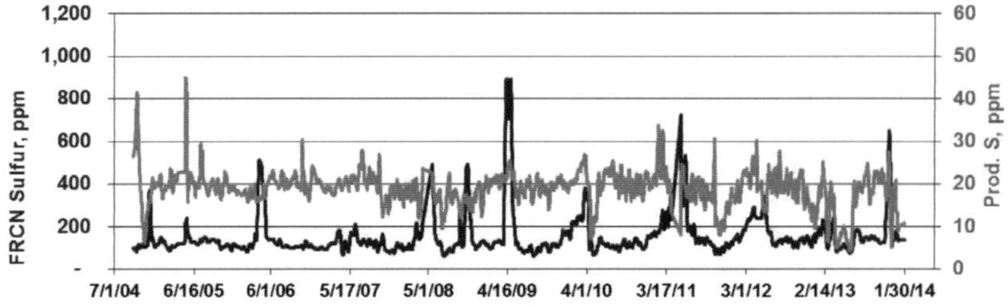

FRCN Processing – Prod S maintained at target value (20 ppm) despite FRCN quality variations

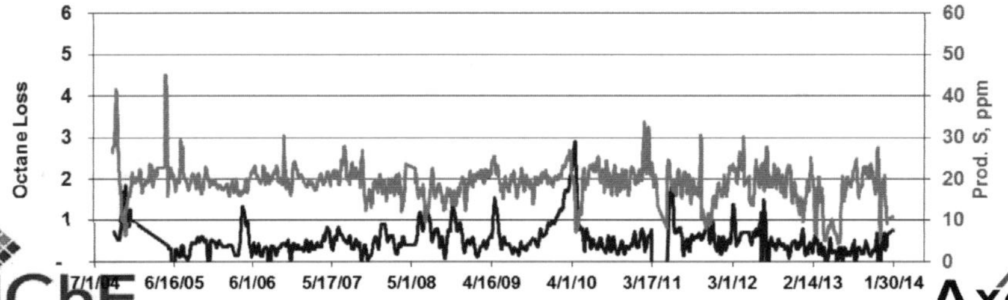

15 AIChE Spring Meeting, Austin, TX

Axens IFP Group Technologies

Conclusion

- Leveraging existing and/or idled assets is critical in a compliance project
- Maintaining flexibility and operability in the post-treater is critical for tight product specs
- With extensive operating experience at 10wppm and a diverse set of processing schemes, Prime-G+ is the technology of choice for Tier 3 compliance

15 AIChE Spring Meeting, Austin, TX

Axens IFP Group Technologies

400303 Szorb Sulfur Removal Technology-a Solution of Tier 3

Abstract:

S Zorb(TM) Sulfur Removal Technology is very different from what is commonly known as the Hydrodesulfurization (HDS) Technologies. What distinguishes S Zorb(TM) SRT from the rest of the commercially available gasoline desulfurization technologies (all are the HDS type) is what matters most to all refiners. As the environmental regulations require lower and lower sulfur content in the gasoline products, most HDS users find themselves needing to spend more capital for upgrading the existing HDS units, taking a hit on lost production due to unit down time, and suffering from additional operating cost due to further loss in Octane Number and liquid yield. The need for more high purity hydrogen would strain further the always tight hydrogen supply situation which is common in most refineries. With S ZorbTM SRT, this can be avoided by designing a unit that will meet your ultimate needs while you operate the unit in an optimal mode for your current needs.

The main chemical reactions in the S Zorb SRT Reactor are sulfur adsorption, olefin hydrogenation, and olefin hydroisomerization. Sorbent regeneration and reduction occur in the Regenerator and Reducer.

Compare to other technologies, SZORB SRT has following advantages:

i. Low octane loss
Our proprietary sorbent helps minimize olefin saturation. By using unique catalytic chemistry, the sulfur is adsorbed from the hydrocarbon rather than released by traditional hydrogenolysis. The sorbent also promotes olefin hydroisomerization, which helps counter octane loss that occurs from olefin hydrogenation. In some cases, it can lead to an overall gain in octane.

ii. Very low volume loss
The S ZorbTM SRT process has very low volumetric losses, and does not trade octane gain for volumetric loss.

iii. Sulfur removal to 10 ppm and lower (from >1000 ppm feed sulfur in one step!)
If your goal is to meet a 30 ppm average product sulfur, our technology can help you achieve this. However, the S ZorbTM SRT process can also help you plan for the future, and for most gasoline feeds, our process can achieve product sulfur levels less then 10 ppm.

iv. Low hydrogen consumption
The S ZorbTM SRT technology has lower hydrogen consumption than hydrotreating, which lowers operating costs and makes hydrogen supplies available for other uses.

v. Ability to use low purity hydrogen
High purity hydrogen is not a requirement for the S ZorbTM SRT process, and our process can operate with low hydrogen purities.

vi. Run lengths to match FCC operations
Because the sorbent catalyst is continuously regenerated, the Reactor activity is stable and does not decrease like fixed-bed, hydrotreating processes. The S ZorbTM SRT process is designed to match the run lengths of FCC units.

vii. Does not require fractionation of FCC stream
If you have an existing fractionator for your FCC naphtha, the S ZorbTM SRT process will treat the heavy or light stream. If you do not have a fractionator, the S ZorbTM SRT will treat the entire naphtha stream. If you have an existing fractionator but would like to employ it for some other use, we can design your unit to treat the combined stream.

viii. Low operating costs
After including costs associated with run length, octane loss, volume loss, and general operating parameters, S ZorbTM SRT technology has comparatively low operating costs.

ix. Competitive capital costs
You might expect a fluidized bed reactor and associated regeneration equipment to add significant capital expenses. Surprisingly, they do not. Conventional metallurgy, low operating pressures, and high space velocities help minimize equipment costs.

Integrating Wastewater Reuse with Seawater Desalination

By William Celenza and Sarah Ou

KBR, Inc.

Houston, Texas

Introduction

Industry accounts for roughly 20% of total freshwater withdrawals globally, with the agriculture and domestic sectors accounting for the remaining 70% and 10%, respectively, as shown in Figure 1. Water use for energy has most often been embedded in "industry". However, the International Energy Agency (IEA) has estimated global water withdrawals for energy production in 2010 accounted for roughly 15% of the world total (IEA), or roughly 75% of all industrial water withdrawals. The Organization for Economic Co-operation and Development (OECD) Environmental Outlook to 2050 predicts that global water demand for manufacturing, not including thermal electricity generation, will increase by 400% from 2000 to 2050 (OECD), the largest increase projected for any sector. Therefore, the 5% of total global freshwater withdrawals by the industry sector in 2000 for manufacturing is projected to increase fourfold by 2050. These figures vary considerably across countries, with more-developed countries having a much larger proportion of freshwater withdrawals for industry than less-developed countries, where agriculture dominates (WWAP, 2014a). For comparison, a recent US DOE analysis (Bauer) of 2005 data, shown in Figure 2, indicated that industry, including thermal electricity generation was 47% of the total US fresh water withdrawals, with the agriculture and domestic sectors accounting for the remaining 40% and 13%, respectively. Note that thermal energy use itself was 40% of US fresh water withdrawals, equaling agriculture. Also in the US, fresh water thermal energy withdrawals in 2005 accounted for only 70% of all thermal energy water withdrawals.

Freshwater shortages are forcing industrial customers to rethink water supply options. The 2014 World Water Assessment Program by UNESCO highlights that abstracting freshwater from a surface or groundwater source, using it, and disposing of it - known as a 'linear approach' - is not sustainable. Future development requires approaches that minimize resource consumption and focus on resource recovery (WWAP, 2014c), versus focusing mainly on compliance with discharge standards. Although using alternative water sources, such as sea or wastewater, can be challenging, they offer a great potential for reducing demands for freshwater (WWAP, 2014b). Many large companies have made considerable progress in evaluating and reducing their water use and that of their supply chains (WWAP, 2015). Most companies have a supply chain water footprint much larger than their operational one; and more than 80% to 90% of the footprint, with most of its water risks, may be beyond its direct operations (Place et al). A water footprint is 'the total volume of water used in the production of the goods and services produced by a business' (WWAP, 2009) and is usually quoted in terms of cubic meters of water per unit product.

Figure 1 Total Global Freshwater Withdrawals

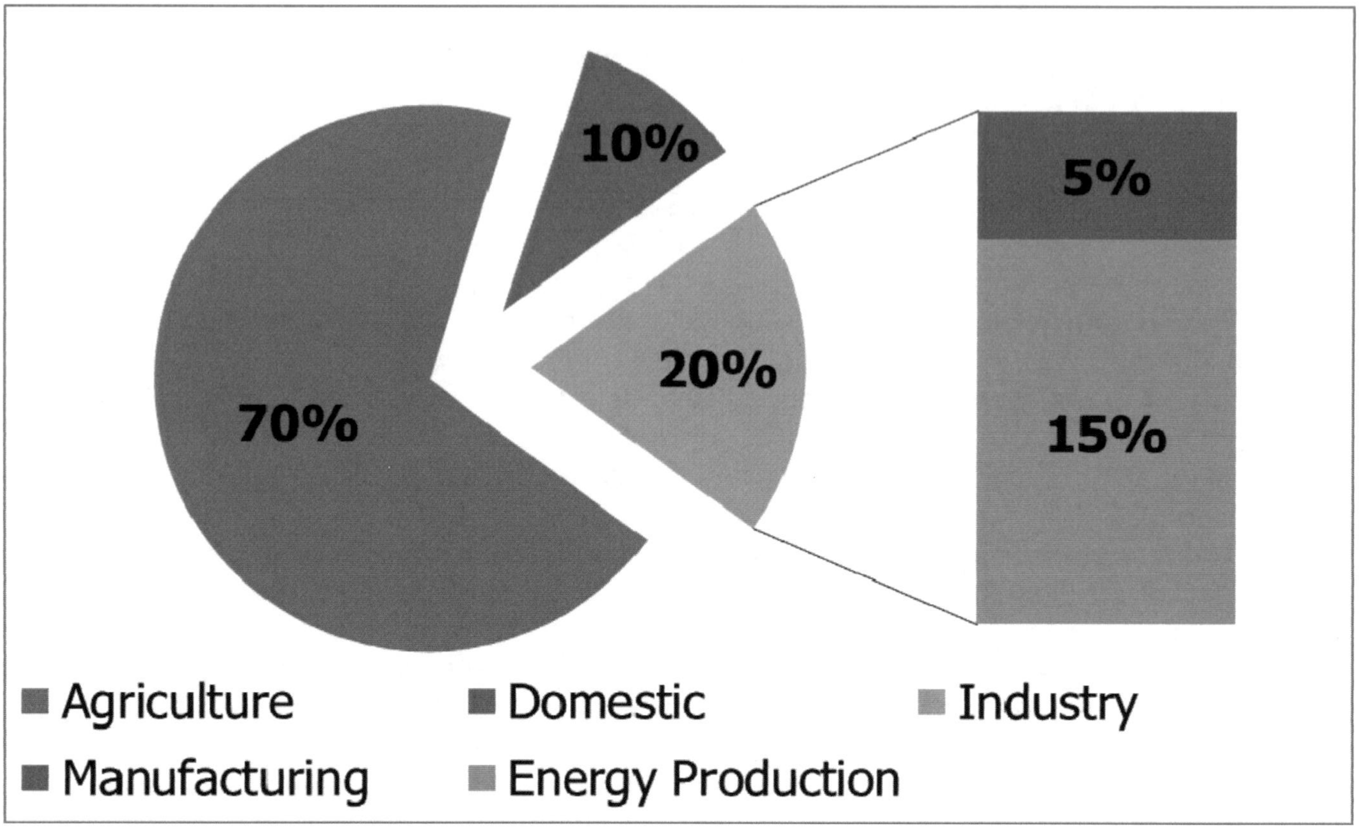

Industrial facilities located near oceans or saline aquifers around the world are relying more on desalination technology to provide freshwater. Industrial facilities are also facing stricter discharge regulations that can favor development of nonconventional water resources, such as wastewater effluent and produced water. Therefore, innovations in technologies that improve wastewater treatment to meet restrictive discharge limits and requirements for reuse, may ultimately lower costs and allow for production expansion without the need to secure more freshwater or increase wastewater treatment capacity.

Seawater desalination and wastewater reuse are now common in coastal regions that experience freshwater shortages. This paper reviews a recent application of desalination and industrial wastewater reuse by reverse osmosis (RO), and describes the methodology used in designing an integrated freshwater production plant and wastewater reuse unit at a large grass root industrial facility in the Middle East. The characteristics of the treated wastewater and reuse requirements and their impact on the overall system are discussed. However, production of potable water by RO is not considered in this paper.

Figure 2 Total US Freshwater Withdrawals

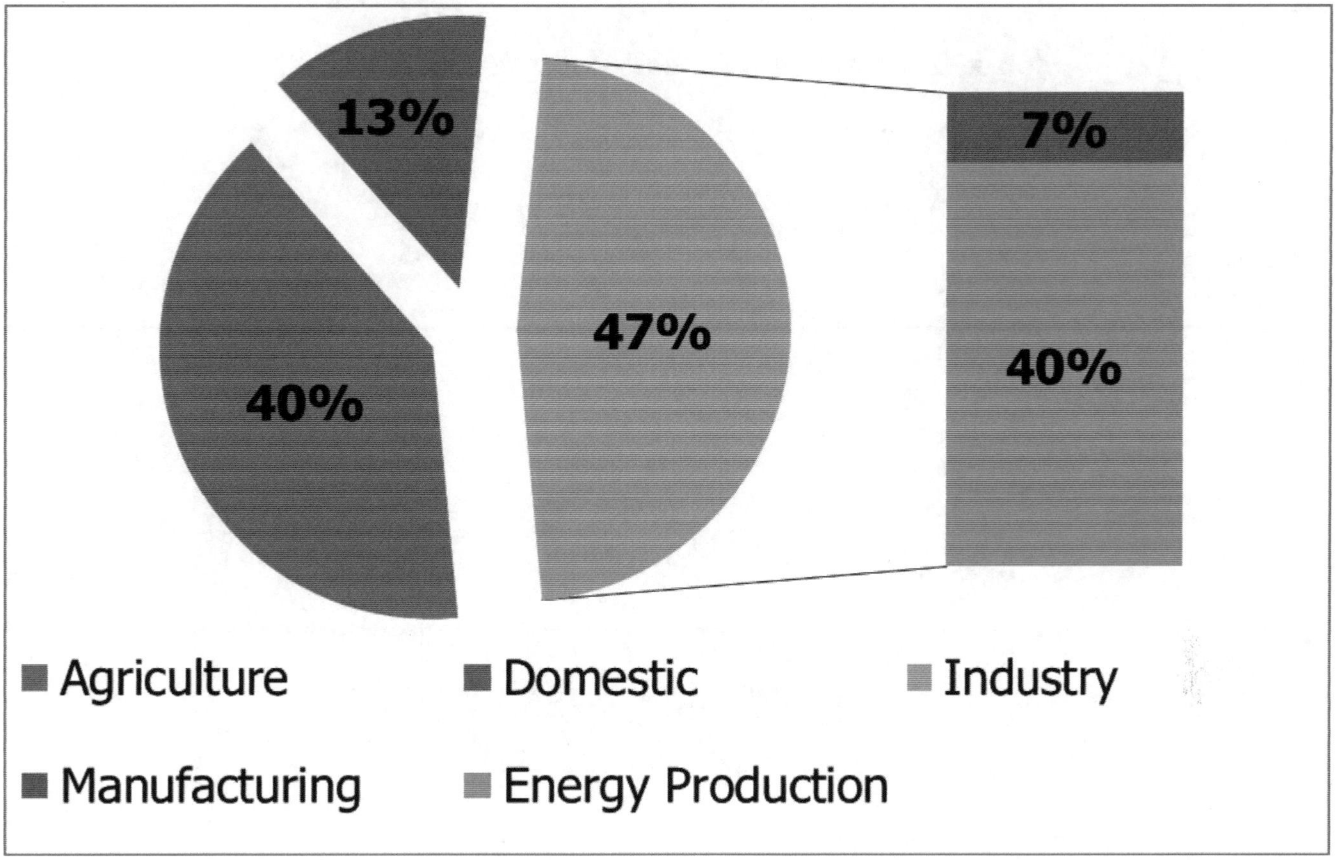

Although the energy costs associated with desalination remain a concern, improvements from advances in system design, energy recovery and operating knowledge have resulted in lower overall costs. The cost for implementing desalination technologies is site specific. For a large desalination plant with a capacity over 100,000 cubic meters per day (m3/day), which is typical for a city or a large industrial application, the cost per cubic meter is inversely proportional to the production capacity. The economic considerations discussed in this paper compare the cost of increasing seawater desalination capacity by RO technology to installing and operating a membrane bioreactor (MBR) based wastewater reuse treatment system capable of producing plant utility water. Included in the cost comparison are a wastewater treatment process designed only for ocean discharge and an upgraded system designed for feed to a post-treatment effluent reuse system, such as an RO, that utilizes a Membrane Biological Reactor (MBR) unit as the biological treatment step.

Integrating Wastewater Reuse into an Industrial facility

Meeting industry's need for water will require an evolution from the common linear approach to water management. Closed-loop systems that recycle water, with a combination of decentralized and centralized elements, can offer superior environmental performance

(Daigger). Also, integrated closed-loop systems as shown in Figure 3, have the potential to satisfy the economic goals of sustainability, as well as environmental goals.

Figure 3 Integrated Closed-Loop Water System

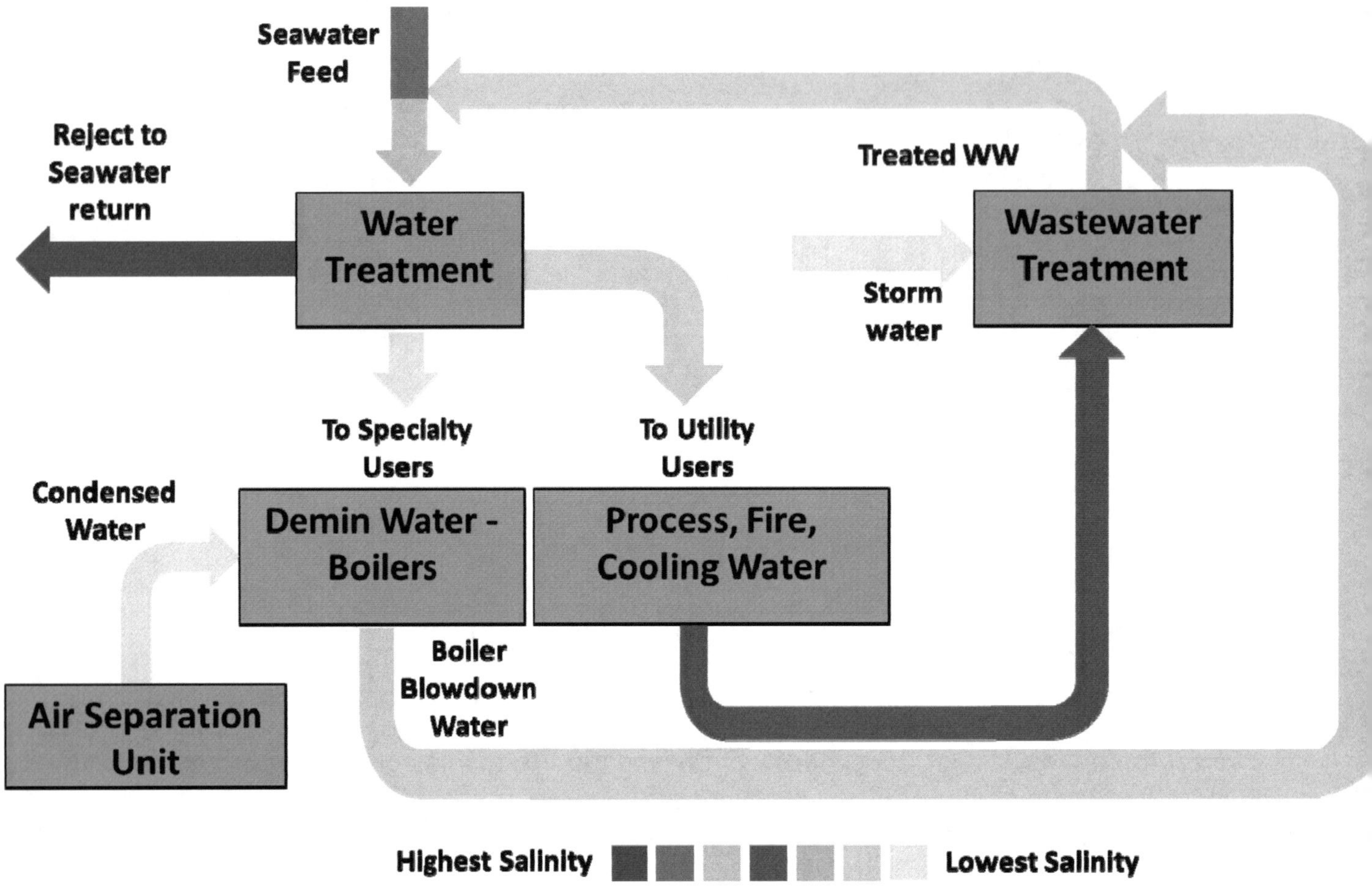

Figure 3 illustrates a closed-loop system for an industrial facility that primarily sources water needs from seawater desalination. Note that the treated water supply is segregated into specialty water and utility water users, such as potable, process, firewater and sanitary supply. Overall, storm water and condensed atmospheric moisture from the site's dedicated Air Separation Unit supply less than 3% of the fresh water needs; however, these water sources offer the lowest Total Dissolved Solids (TDS) options available.

Industry has typically viewed wastewater as more of a compliance issue than a strategic resource. The end-point of conventional wastewater treatment has been the production of a treated wastewater effluent suitable for discharge to the environment (Tchobanoglous). The main advantage of wastewater reuse in Figure 3 is a direct offset of the amount of seawater needed to be desalinated, because the total dissolved solids content of wastewater (1,000 mg/L) is much smaller than of seawater (35,000 to 45,000 mg/L). However, a Wastewater Treatment Plant (WWTP) in a reuse scheme is subject to a two-tiered treatment specification requirement. First the treated effluent quality must allow for direct use as feed to other technologies, such as a Reverse Osmosis (RO) Unit. The other criteria are the "discharge to coastal water" specific performance standards. Thus the WWTP effluent must meet the most restrictive of these requirements. In the event of a treatment upset at the WWTP, depending on the extent of disruption and the duration, the WWTP effluent may need to be detained onsite or discharged to the seawater outfall if circumstances permit. Also a process upset at a WWTP in a reuse scheme has a direct economic impact on water sourcing costs as well as increased treatment and environmental compliance costs. As such, a more robust or redundant treatment system than a discharge only version should be considered and additional discipline on managing a site sewer system would be warranted. For instance, streams that are mostly oil should be collected separately from the oily water sewer system and managed outside the WWTP. Also, the repeated recycling of water through an RO unit may result in the buildup of trace constituents, such as silica, some Volatile Organic Compounds and Boron which must be managed to maintain the quality of the water for its intended uses.

Design Considerations for Integrating Wastewater Reuse with Seawater Reverse Osmosis (RO) Desalination

The quality of feed water is important to RO system design. A typical One-Pass RO desalination unit, shown in Figure 4, consists of three main sections: Feed and Pretreatment, RO membranes and Permeate Post-Treatment. The requirements of each section vary per the source water quality. For example, the chemical characteristics of Red Sea surface water, Arabian Gulf, and Mediterranean Seawater impose additional load on an RO system due to elevated levels of TDS above standard seawater composition (DOW), over 40,000 mg/l versus 35,000 mg/l. Theoretically, the required applied pressure to overcome the natural osmotic pressure is about 0.78 bar/1000 mg/L of TDS (Hafez et al). Figure 5 shows typical RO feed pressures for an elevated TDS desalination application are over threefold that of a wastewater reuse system, 70 versus 20 barg.

Also desalination is impacted by temperature, with a high average seasonal temperature allowing for a lower operating pressure but decreasing permeate quality. However, since a wastewater treatment system temperature is limited to biological treatment system requirements, a cooling step would be required as part of a reuse system for source temperatures above 32 C. Likewise, even though wastewater effluent is typically injected with chlorine for disinfection, a downstream dechlorination unit may be needed due to limited RO membrane tolerance to chlorine.

The complexity of an RO system is also determined by the desired product quality. In Figure 5, potable water and utility water, are produced in the 1st-pass, while higher quality water suitable for feed to a demineralizer is produced after a 2nd-pass. Normally, potable water and utility water are of the similar quality. If desired product from the RO system is high purity water for processing through a demineralizer, which is higher quality water than utility water, two passes of RO are needed. The first pass generates utility water or potable water and the second pass generates high purity water.

Figure 4 One Pass RO

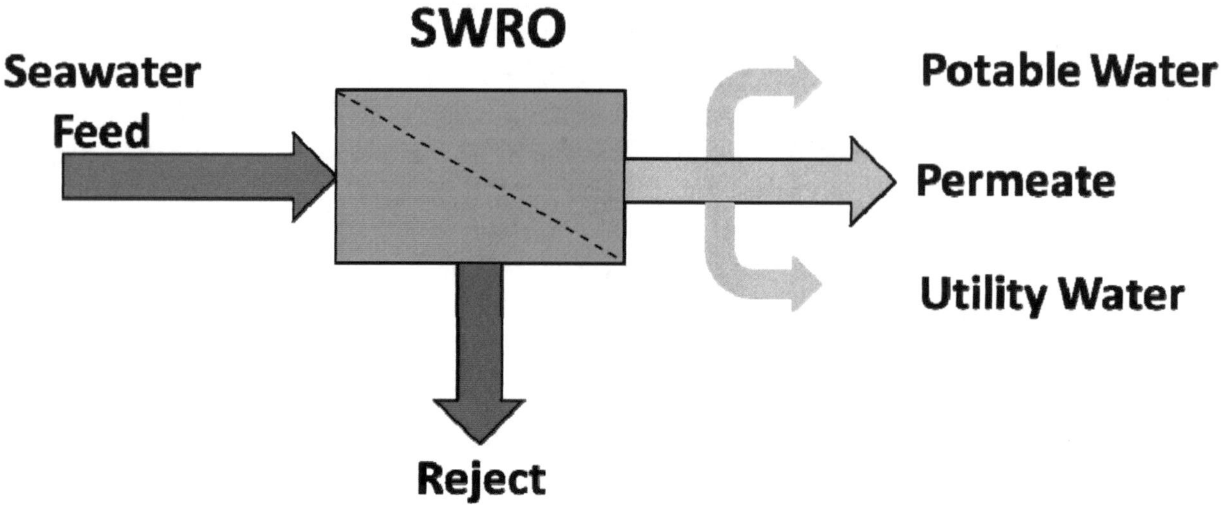

Figure 5 Integrated Seawater RO and Wastewater RO

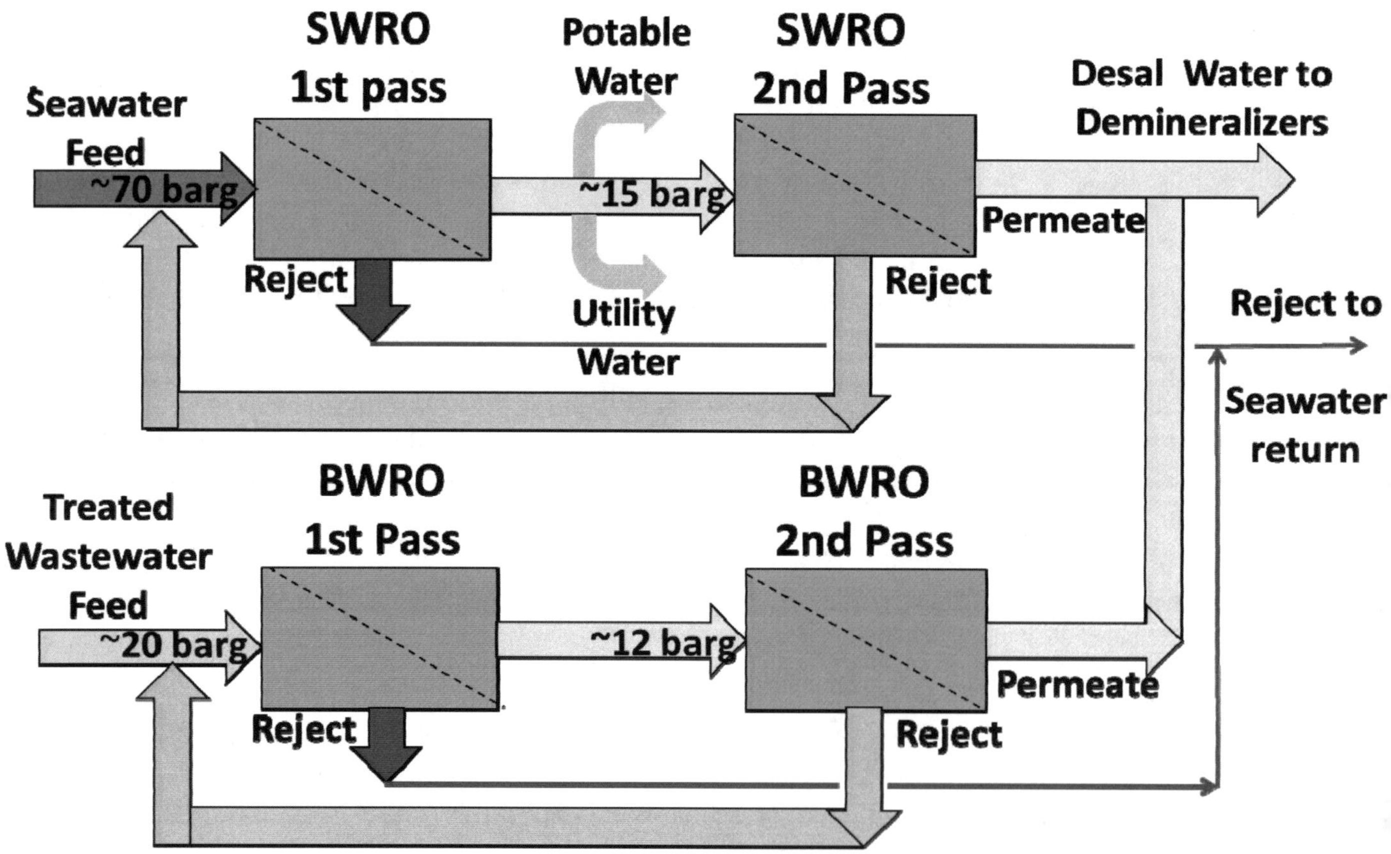

Finally as shown in Figure 5, the RO unit 1st pass generates reject brine which is typically suitable for return discharge to the seawater source. However, a wastewater reuse brackish water RO (BWRO) reject brine may exceed discharge standards due to an up to 3:1 concentration factor of biotreatment effluent. For instance if the discharge option crtieria is 25 mg/l for BOD, then the biotreatment step would need to reduce BOD to 6 mg/l to compensate for the concentration factor. Thus, if dilution with a seawater RO (SWRO) reject brine is not an option, additional post treatment or an even lower biotreatment effluent specification would be required. The above considerations are summarized in Table 1.

Table 1 – Design Considerations for Seawater vs Wastewater Reuse

Item	Seawater	Wastewater Reuse
TDS Level	Locally elevated TDS increases required operating pressure.	Not directly impacted
Temperature	Lower operating pressure but decreased permeate quality.	Effect is limited due to max Biotreatment operating requirements of 32 C. However, a cooling step would be required if over.
Disinfection	Chlorine injection may be required but dechlorination typically required due to impact on RO membranes.	Same as for Seawater.
Product Quality	One or more RO passes to meet quality specification.	Same as for Seawater
Reject brine disposal	Usually direct return discharge to seawater source is permitted.	Concentration of wastewater constituents in reject brine may exceed discharge standards. Dilution with seawater reject brine or additional treatment would be required.

Figure 6 shows a diagram of a typical integrated RO desalination package. The recycling of the second pass RO reject, or concentrate, increases the overall recovery. Also, the cleaning and backwash streams are also collected and treated for reuse. Only one stream, the neutralization wastewater from demineralizer and condensate polisher regeneration cycles is not collected for reuse since this stream was considered a high operational risk due to high solids and scaling potential. All other wastewater streams, such as the higher quality boiler blowdown and the treated wastewater are blended to dilute constituents from both streams, additives from the boiler blowdown and residual organics from the treated wastewater.

Figure 6 Typical RO Desalination Package

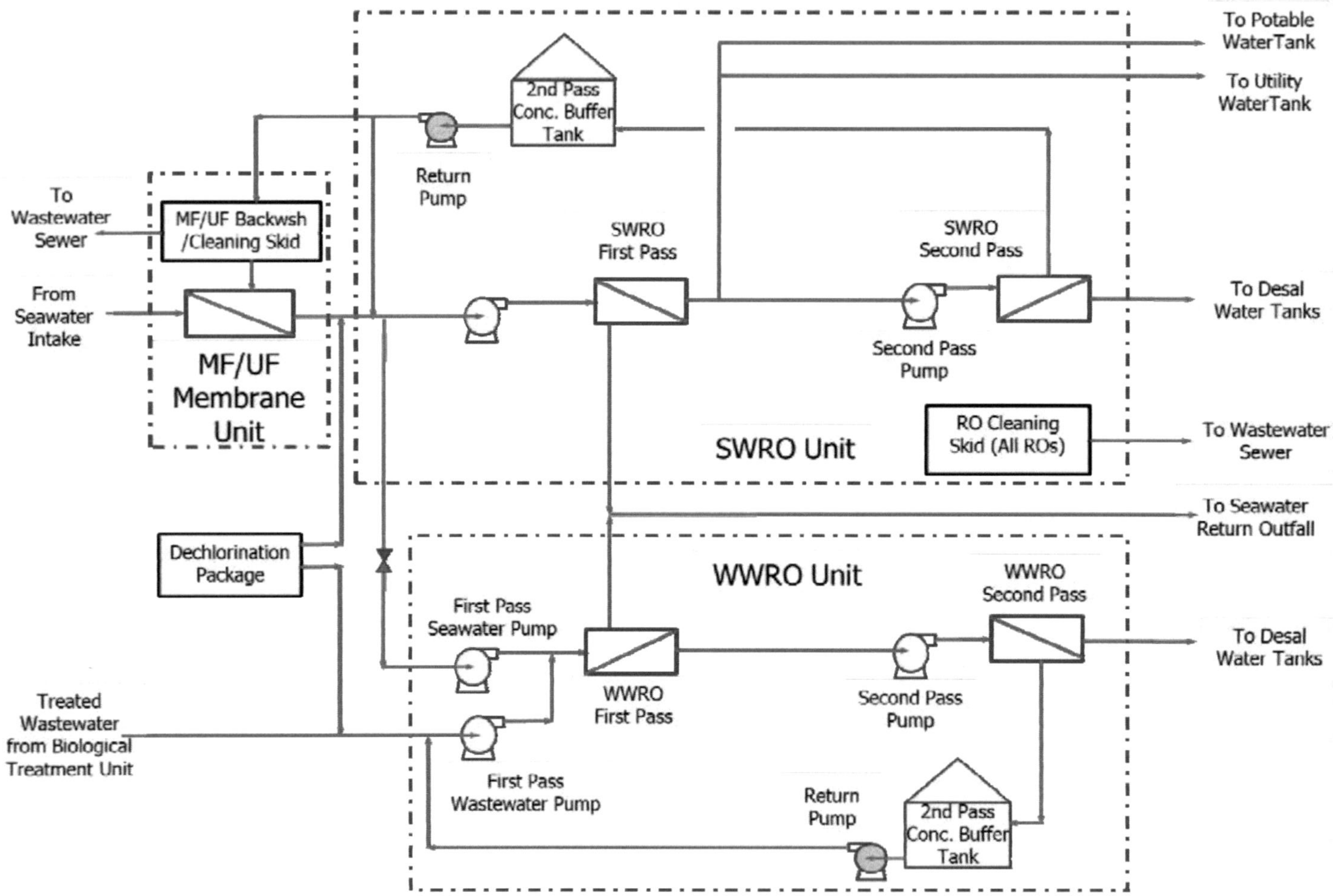

Thermoeconomic Analysis of Desalination

The energy cost of treating low salinity seawater is approximately 10 times that of a typical freshwater source; thus the desalination of seawater should only be considered when there are no other sources or the cost of transporting water is very high (Pearce et al). Likewise, approximately twice as much energy is required to deliver treated water from wastewater as from a lake or river. For industry, the implication is that in terms of energy required, it may be more economical to extract water directly from a raw external source or treat wastewater to minimum standards for discharge, than to fully treat wastewater for reuse (WWAP, 2014a and Scientific American).

Thermoeconomics uses economic principles to evaluate the interaction between cost and efficiency, by applying thermodynamics as a basis to assign costs. For the energy intensive desalination process, thermoeconomics provide a method for conducting feasibility and optimization studies during a design phase. The "thermodynamic minimum energy" for desalination is the same for all processes, with the energy differences between different process technologies due to the difference in process efficiency. For competing desalination processes, RO and Mechanical Vapor Compression (MVC), which get their energy from electricity, have better economics and thermodynamic performance than Multistage Flash (MSF), Multi-Effect Evaporation (MEE) and Thermal Vapor Compression (TVC), which are thermal processes.

To evaluate the process efficiency benefit of integrating wastewater reuse with seawater desalination, the minimum separation work requirement for incomplete separation of saline water was calculated (Cerci et al). The input parameters for an all seawater feed case, and an integrated treated wastewater and seawater case are shown in Table 2.

Table 2 – Minimum work of separation for saline water at 34 C

Type of Saline Water	Salinity %	Recovery %	Product Water TDS mg/l	Minimum work of separation, kJ/kg Product Water	Minimum power input, kWh/m^3 Product Water
Red Sea Seawater	4.6	33	105	4.0	1.1
Red Sea Seawater 3:1 with Treated wastewater	3.5	44	105	3.1	0.86

The minimum work values in Table 2 indicate a 22% decrease in energy required for the integrated treated wastewater and seawater case. This comparison is expected since the treatment cost decreases as the quality of the source water increases. In general, the amount of energy needed in water and wastewater treatment processes varies greatly and is dependent upon factors such as the quality of the source water, the nature of any contamination, and the types of treatment used by the facility as shown in Table 3. In practice, more sophisticated techniques, such as reverse osmosis, require relatively larger energy inputs (1.5-3.5 kWh/m^3) (WWAP, 2014a). The energy demand of the currently best available seawater RO desalination process including pumping, pre-treatment, RO desalting, and post-treatment is typically 3.4 to 4.8 kWh/m^3 (Siemens), with boron treatment the energy demand may approach 7 kWh/m^3 (DESWARE). For comparison, even an efficient desalination plant such as the Ashkelon plant in Israel, which requires 2.9 kWh/m^3, is more energy intensive than the transportation of water over long distances in California, which requires 2.5 kWh/m^3 (WWAP, 2014c). However, the desalination industry is working on reducing energy costs. While energy consumption has been significantly reduced in the past two decades, the International Desalination Association's (IDA's) Energy Task Force has specifically challenged the desalination industry to achieve a further reduction of 20% in energy requirements for seawater desalination by 2015 (IDA).

Table 3 Energy Required to Provide Water Safe for Human Consumption

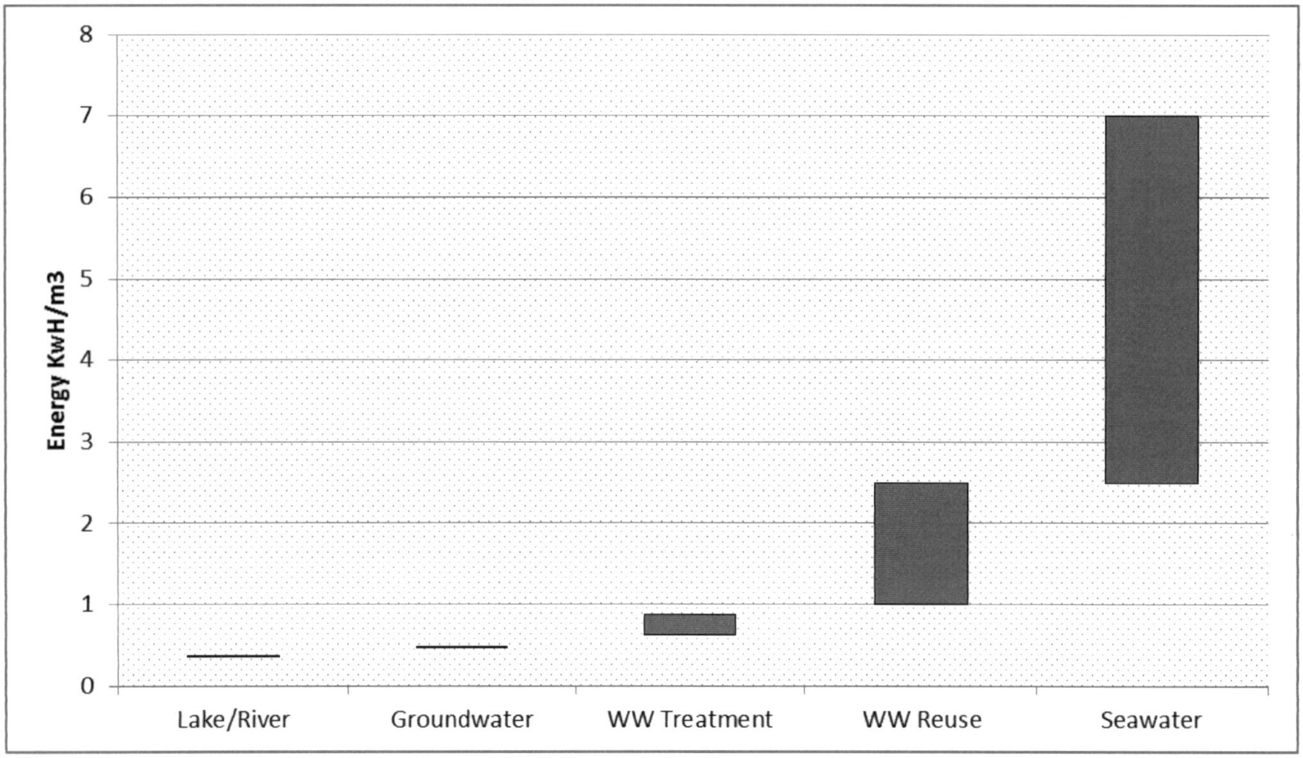

Sources: Amended Diagram based on WWAP 2014a, WBSCD, Scientific American

Concluding Thoughts

Global economic growth is increasingly coming up against environmental restraints. These constraints include water scarcity, and thus the desalination and water reuse industry has emerged as important in achieving water efficiency. Technologies associated with salt removal and the reuse of wastewater will increasingly become an essential part of sustainable economic growth and profitability. The ability to turn low quality wastewater and raw water sources, such as seawater, into high quality process water, has been key to enabling the expansion of the downstream oil and gas industry in water short areas such as the Middle East.

References

Bauer, D. Water-Energy Nexus: Challenges and Opportunities, AIChE Webinar Series. Water Withdrawals and Thermoelectric Generation slide. Office of Energy Policy and Systems Analysis, US Department of Energy (DOE). March 25, 2015.

Cerci, Y, Y Cengel, B Wood, N Kahraman, ES Karakas. Improving the Thermodynamic and Economic Efficiencies of Desalination Plants: Minimum Word Required for Desalination and Case Studies of Four Working Plants. US Department of the Interior, Bureau of Reclamation. November 2003.

Daigger, G.T. New Approaches and Technologies for Wastewater Management. The Bridge (National Academy of Engineering, Washington DC), Vol 38 No 3 fall 2008, pp 38-45

DESWARE, Encyclopedia of Desalination and Water Resources, Table 1. www.desware.net/desa4.aspx accessed on March 24, 2015

DOW, Tech Manual Excerpt Form No. 609-02010-1004. Table 2.1

Hafez, A, S Al-Manharawy. Economics of seawater RO desalination in the Red Sea region, Egypt. Part 1 A case study. (2002) Desalination 153, pp 335-347.

IDA (International Desalination Association) DESALINATION – AN OVERVIEW Updated: November 2013, accessed on March 23rd 2015. http://idadesal.org/desalination-101/desalination-overview/

IEA (International Energy Agency). 2012. World Energy Outlook 2012. Paris, OECD/IEA.

OECD (Organization for Economic Co-operation and Development). 2012. *Environmental Outlook to 2050: The Consequences of Inaction, Key Facts and Figures*. Paris, OECD. http://dx.doi.org/10.1787/9789264122246-en

Pearce, G.K. 2012. *Desalination vs water reuse: An energy analysis illustrated by case studies in Los Angeles and London*.V. Lazarova, K. Choo and P. Cornel (eds), 2012: Water-Energy Interactions in Water Reuse. London, IWA Publishing, p 265 Chapter 18.

Place J., Dutto, P.R. and Casula, V. 2012. Putting water in the mainstream of your business strategy. Prism 1: 69-81. London, Arthur D. Little. http://www.adlittle.com/downloads/tx_adlprism/Prism_01-12_water.pdf

Scientific American, Earth3.0, Special Issue, "Energy vs. Water. Why both crises must be solved together", October 2008, with amended data from GHD.

Siemens Press Release. Singapore July 4, 2011 SIWW. Industry Sector, Industry Solutions Division. Reference number: IIS201106.205s fp.

Tchobanoglous, G, H.D. Stensel. 2014.The Future of Activated Sludge. Environmental Engineer & Scientist, Winter 2014 V 50 N 1. pp 47-49

WBCSD (World Business Council on Sustainable Development). 2009. *Water, Energy and Climate Change: A Contribution from the Business Community*. Geneva, WBCSD. http://www.wbcsd.org/pages/edocument/edocumentdetails.aspx?id=40&nosearchcontextkey=true

WWAP (World Water Assessment Programme). 2009. *The United Nations World Water Development Report 3: Water in a Changing World*. Paris/London, UNESCO/ Earthscan. p. 101

WWAP (World Water Assessment Programme). 2014. *The United Nations World Water Development Report 2014: Water and Energy*. Paris, UNESCO.

a) Chapter 2 Water: Demands, energy requirements and availability
b) Chapter 5 Infrastructure
c) Chapter 7 Cities

WWAP (United Nations World Water Assessment Programme). 2015. *The United Nations World Water Development Report 2015: Water for a Sustainable World*. Paris, UNESCO. Chapter 9 Industry.

Fundamentals of Refinery Wastewater Treatment system

Rafique Janjua PE
Fluor Enterprise Inc
Houston, Texas

15 AIChE Spring Meeting, Austin, TX

FLUOR®

Sources of Refinery Wastewater

Utility Wastewater:

- Cooling Tower blow down
- Boiler blow down
- Once through service water
- Pump seals flush water
- Wastewater from BFW make up system
- Filter backwash water
- Wastewater from raw water pretreatment systems
- Contaminated storm water
- Potentially contaminated storm water

15 AIChE Spring Meeting, Austin, TX

FLUOR®

Process wastewater:

- Crude desalting
- Distillation
- Fluid Catalytic Cracking
- Catalytic Reforming
- Alkylation
- Thermal cracking/Visbreaking
- Catalytic Hydrocracking/ Hydrotreating
- Cocking
- Isomerization
- Sulfur Removal/Claus process
- Sweetening/Merox process

15 AIChE
Spring Meeting, Austin, TX

FLUOR®

Wastewater Minimization, Classification, Collection and Treatment

15 AIChE
Spring Meeting, Austin, TX

FLUOR®

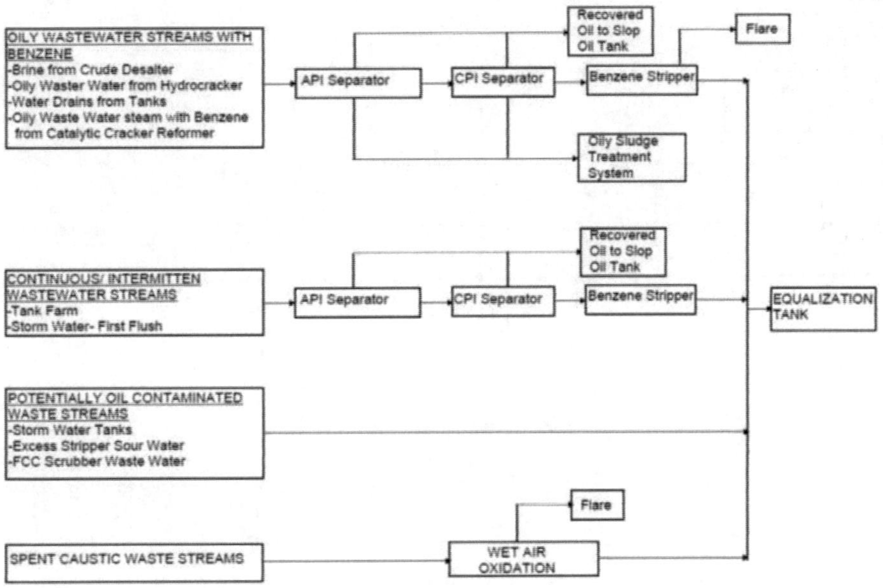

15 AIChE Spring Meeting, Austin, TX

FLUOR®

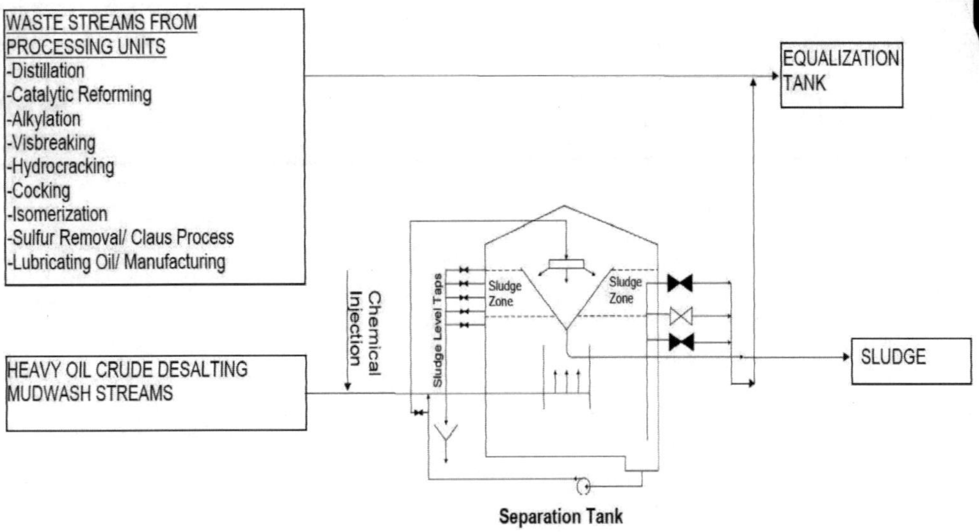

15 AIChE Spring Meeting, Austin, TX

FLUOR®

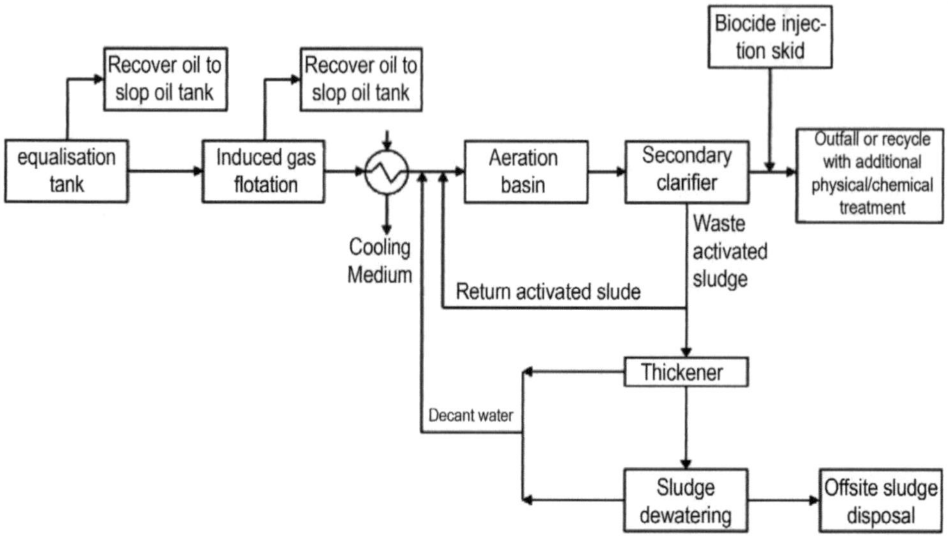

15 AIChE Spring Meeting, Austin, TX

FLUOR®

QUESTIONS ?

15 AIChE Spring Meeting, Austin, TX

FLUOR®

CFD Modeling of Air Emissions from Controlled Flare Tests

Vijaya Damodara, Lan Liu, Kader Rasel, Ajit Patki, Hashim Almrayani, Daniel Chen, Helen Lou, Xianchang Li, Peyton Richmond

CFD Lab, Lamar University, Beaumont, TX, USA

Kanwar Singh, CBI Engineering, Baton Rouge, LA, USA

Matthew Johnson, Energy & Emissions Lab, Carleton University, Ottawa, Ontario, Canada

15 AIChE
Spring Meeting, Austin, TX

Contents

- ➢Introduction
- ➢Contents

- ➢Part-1 Reduced mechanism
- ➢Methodology
- ➢Results and discussion

- ➢Part-II CFD simulations overview
- ➢Methodology
- ➢Results and discussion

- ➢Conclusion

15 AIChE
Spring Meeting, Austin, TX

Introduction

- Flare emissions account for 60% of the Highly Reactive Volatile Organic Compounds (HRVOCs) based on the 2007 HRVOC special emissions inventory
- Flare emissions depend heavily on a flare's combustion efficiency
- Readily available combustion mechanisms are quite large and involve 1000's of reactions and 100's of species
- Recent studies/gas imaging IR camera suggest that flares may not be as efficient as claimed[1-2]
- New reduced combustion mechanism, LU 3.0.1 developed by Lamar research group

Baukal, Charles E., Schwartz, Robert E., The John Zink Combustion Handbook; ISBN 0-8493-2337-1, CRC Press: Boca Raton, 2001.
Technical Supplement 4: Flares. TCEQ Publication RG-360A, January 2007.

15 AIChE Spring Meeting, Austin, TX

Flare Emission Species

- Rigorous combustion mechanisms suggest numerous incomplete combustion species are produced in combustion reactions
- These include radicals (HO, & HO_2), radical producing species (CH_2O, C_2H_4O, HONO), NO_x, CO, and soot [6,7,8]
- In recent past, Lamar research team has simulated several laboratory flames (e.g., Sandia flame and McKenna flat flame) with good agreements
- Soot or visible smoke, is of particular concern for flare operators / regulatory agencies

6. *Michael Frenklach, "Reaction mechanism of soot formation in flames," Phys. Chem. Chem. Phys., 2002,* ***4, 2028–2037.***
7. *Davis, S. G.; Law, C. K. Determination of and Fuel Structure Effects on Laminar Flame Speeds of C1 to C8 Hydrocarbons, Combustion Science and Technology, 1998, 140(1), 427- 449.*
8. http://ignis.usc.edu/Mechanisms/C3/c3.html

15 AIChE Spring Meeting, Austin, TX

Full/Reduced Combustion Mechanisms

- GRI-Mech 3.0 - (53 species and 325 reactions)
- USC-II - 111 species and 784 reactions. Based on GRI -3.0
- Lamar research group developed 50-species reduced mechanisms for the combustion of C_1-C_3 light hydrocarbons (LU1.0, 1.1, and 2.0 based on GRI-3.0 and USC-I mechanisms)
- Successfully validated against experimental data in CHEMKIN-PRO
- Tested in simulations for its further validation
- LU 3.0.1 - Capable of predicting soot in TCEQ air cases in CFD simulations

Wang, H.; Laskin, A. A Comprehensive Kinetic Model of Ethylene and Acetylene Oxidation at High Temperatures, University of Southern California, Progress Report, http://ame-www.usc.edu/research/combustion/combustionkinetics/Mechanisms/C2-C4/c2.html
http://ame-www.usc.edu/research/combustion/combustionkinetics/Mechanisms/USC-Mech%20II/USC_Mech%20II.htm
Reaction Workbench 15131,
Reaction Design: San Diego, 2013.
CHEMKIN 10131, Reaction Design: San Diego, 2013

15AIChE
Spring Meeting, Austin, TX

50-Species Mechanisms

LU 1.1 (With NO_2)

- In 2010 John Zink flare tests, NO_2 is a monitored species.
- The full 93 species mechanism was reduced based on mole fractions, rate constants, & their effect on other major species for use in the EDC model.
- CN was replaced with NO_2 in LU 1.1.
- Runs more efficiently (in-house UDF available)

LU 3.0.1 (with soot precursors)

- Attempts to strike a balance between Volatile Organic Compounds (VOCs) emissions and soot emissions
- New mechanism with 50 species and 310 reactions has been compared to the full USC II mechanism
- Contains important soot precursor species and can handle C_1-C_4 hydrocarbons
- n-butane, 1-butene and 1,3-butadiene included

15AIChE
Spring Meeting, Austin, TX

Important species comparison between various mechanisms

Comparison reduced mechanisms for mole fraction of major species at residence time of 1 sec for C3H6 fuel

Species	USC II	LU 3.0.1	Abs. error %	LU 1.0	Abs. error %	LU 2.0	Abs. error %
C2H2	1.46E-06	1.32E-06	9.72	1.19E-06	18.53	1.21E-06	17.23
CH4	3.01E-07	3.02E-07	0.23	3.30E-07	9.68	3.91E-07	30.07
CO	5.59E-03	5.70E-03	1.97	6.24E-03	11.68	5.98E-03	7.06
CO2	1.21E-01	1.21E-01	0.2	1.20E-01	0.64	1.21E-01	0.35
H2	1.35E-03	1.37E-03	2.1	1.51E-03	12.31	1.45E-03	7.41
Average abs. error %			2.84		10.57		12.42

Comparison of errors for mole fraction of trace species at residence time of 1 sec for C_3H_6 fuel

Species	USC II	New	Error, a factor of	LU 1.0	Error, a factor of	LU 2.0	Error, a factor of
C3H6	4.35E-08	4.67E-08	1.07	4.59E-08	1.05	1.48E-07	3.41
HO2	1.19E-07	1.16E-07	1.02	2.06E-07	1.73	2.51E-07	2.1
OH	1.27E-03	1.33E-03	1.05	1.26E-03	1	1.22E-03	1.04
CH3CHO	3.16E-12	4.22E-12	1.34	3.58E-11	11.34	4.44E-13	7.11
CH2O	2.11E-07	1.68E-07	1.25	1.75E-07	1.21	2.02E-07	1.04
C2H4	1.81E-08	3.16E-08	1.74	3.84E-08	2.12	2.47E-08	1.36
CH2CHO	8.80E-12	1.10E-11	1.25	1.07E-11	1.21	1.70E-11	1.93
Average error		1.25		2.81		2.57	

15 AIChE Spring Meeting, Austin, TX

Species Selection for Developing a Reduced Mechanism

- Mechanisms used: LU1.1 and LU2.0* mechanisms & USC II
- Common VOC species unchanged in new mechanism
- New reduced combustion mechanism with 50 species and 310 reactions was developed from the USC-II mechanism
- Contains important soot precursor species and can handle combustion of C_1-C_4 hydrocarbons (n-butane, 1-butene and 1,3-butadiene included)
- Predicts both volatile organic compounds (VOCs) and soot emissions
- New mechanism has been compared to the full USC II mechanism and successfully validated against experimental data in CHEMKIN-PRO

*Helen H. Lou, Christopher B. Martin, Daniel Chen, Xianchang Li, Kyuen Li, Hitesh Vaid, Anjan Tula Kumar, Kanwar Devesh Singh, & Doyle P. Bean, "A reduced reaction mechanism for the simulation in ethylene flare combustion," Clean Technologies and Environmental Policy, **14** (2), pp 229-239, April 2012, doi:10.1007/s10098-011-0394-9

15 AIChE Spring Meeting, Austin, TX

Reaction Path Diagram (I)

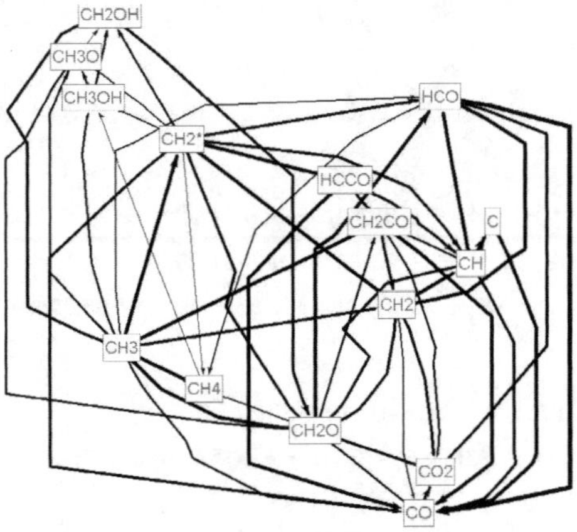

CH4-CO2 with 15 species

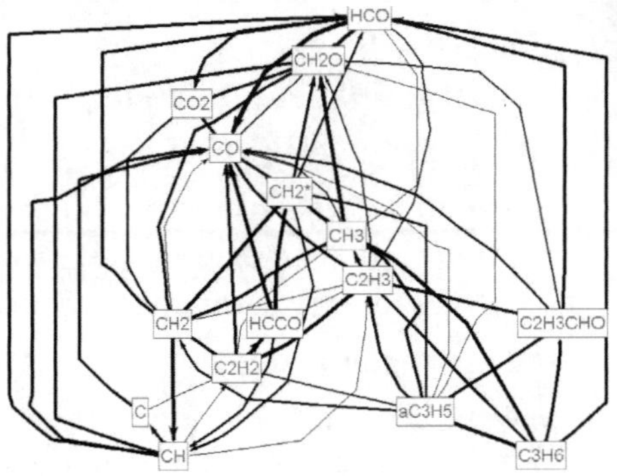

C3H6-CO2 with 15 species

15 AIChE
Spring Meeting, Austin, TX

Reaction path diagram(II)

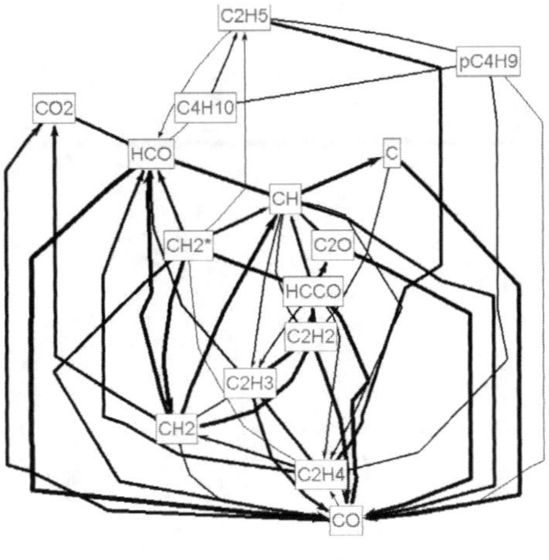

C4H10-CO2 with 15 species

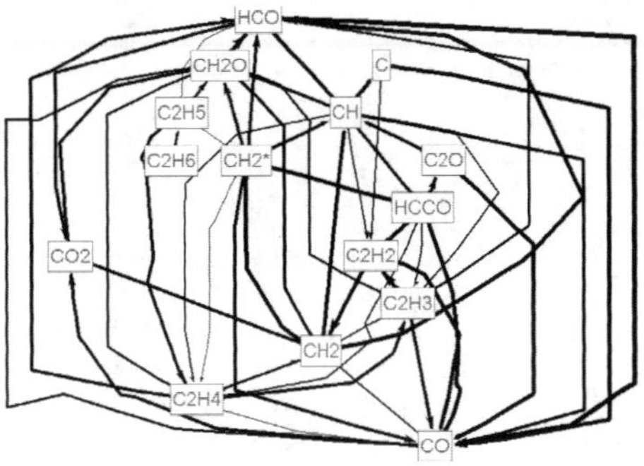

C2H6-CO2 with 15 species

15 AIChE
Spring Meeting, Austin, TX

Validation: Key Performance Indicators of a Flame

- Laminar flame speed*
  - -Speed at which a laminar flame propagates through a pre-mixture of fuel and air
  - -Strongly influences design parameters of combustion equipment.
- The ignition delay time*
  - -Period between the creation of a combustible mixture when the fuel is injected in an oxidizing environment, and the sustained, on onset of the rapid reaction phase leads to the rise of temperature and pressure
- Adiabatic flame temperature*
  - -Measure of the maximum temperature that could be reached by combusting a particular gas mixture under a specific set of conditions

*C.K. Law, A. Makino, T.F. Lu, On the off-stoichiometric peaking of adiabatic flame temperature, *Combustion and Flame, Volume 145, Issue 4, June 2006, Pages 808-819, ISSN 0010-2180,*
**Vagelopoulos, C.M., Egolfopoulos, F.N. Law, C.K. Proc. Combust. Inst. 1994, 25, 1341*
**Jomaas, G., Zheng, X. L., Zhu, D. L., and Law, C. K. Proc. Combust. Inst. 2005, 30, 193.*

15 AIChE
Spring Meeting, Austin, TX

Laminar Flame Speed

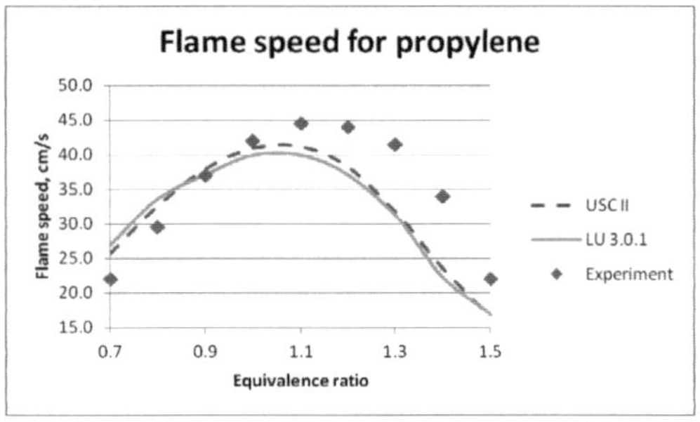

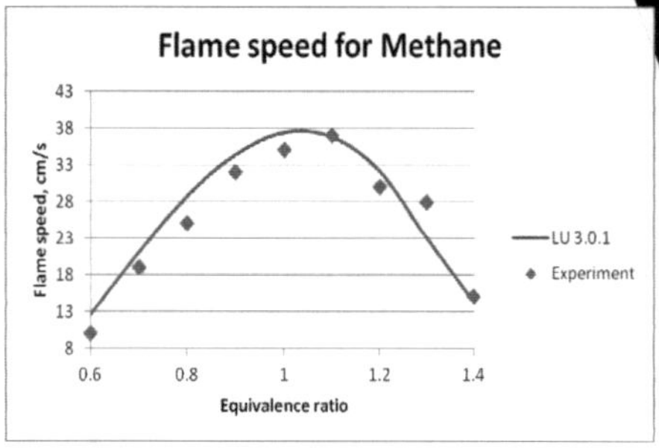

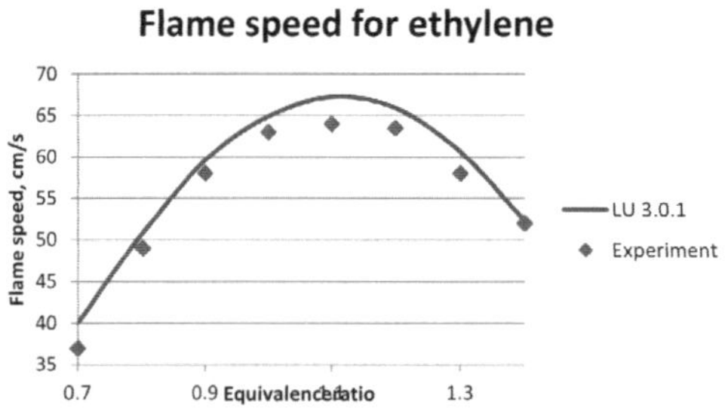

15 AIChE
Spring Meeting, Austin, TX

Ignition Delay and Adiabatic Flame Temperature

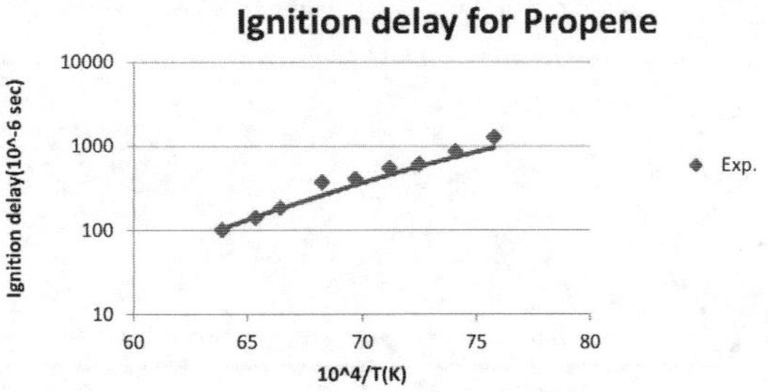

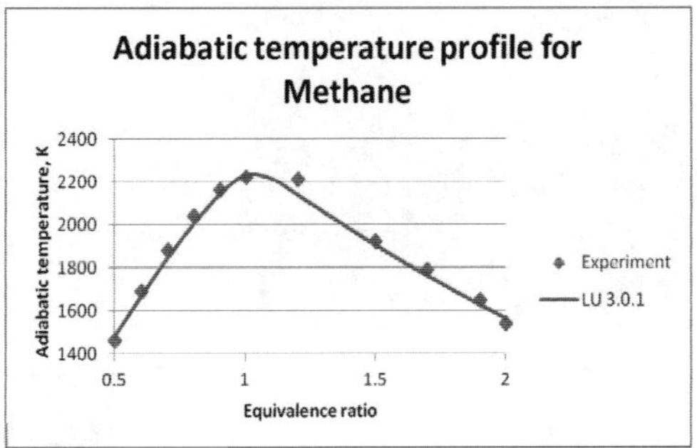

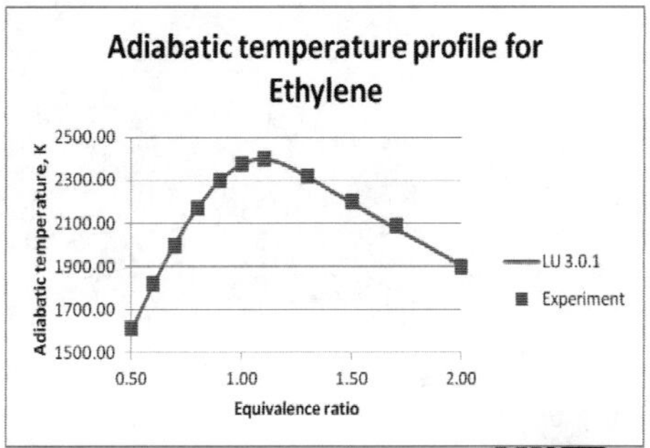

15 AIChE Spring Meeting, Austin, TX

Need for CFD Modeling

- Flare field tests are expensive
- Validated, rigorous CFD flare models are useful for parametric studies & flare optimization
- Modeling includes:
  1. Geometry and domain
  2. Coordinates
  3. Governing equations
  4. Flow conditions
  5. Initial and boundary conditions
  6. Selection of models for different applications

15 AIChE Spring Meeting, Austin, TX

Air-Assisted Flare: Sep. 2010 TCEQ/John Zink Flare Campaign, Tulsa, OK

Elevated Position

2010 TCEQ Flare Study Final Report, UT/CEER, TCEQ PGA No. 582-8-86245-FY09-04 and Task Order No. UTA10-000924-LOAT-RP9, Aug. 1, 2011.

15 AIChE
Spring Meeting, Austin, TX

Air assisted flare stack and flare tip

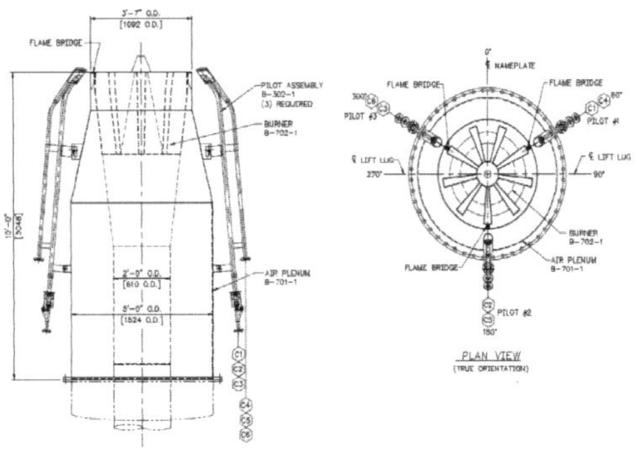

Terminology

Destruction & Removal Efficiency (DRE)

$$\text{DRE} = \frac{(C_xH_y \text{ fed} - \text{unburned } C_xH_y) \times 100\%}{C_xH_y \text{ fed}}$$

Combustion Efficiency (CE)

$$\text{CE} = \frac{(\text{Carbon converted to } CO_2) \times 100\%}{(\text{Carbon fed as fuel})}$$

When determined in the field, CE is calculated based on concentration ratios:

$$\text{CE (\%)} = \frac{CO_2 \text{ (plume)} \times 100}{CO_2 \text{ (plume)} + CO \text{ (plume)} + \sum \text{hydrocarbons (plume)} + \text{soot}}$$

15 AIChE
Spring Meeting, Austin, TX

Geometry and Meshing for CFD Analysis

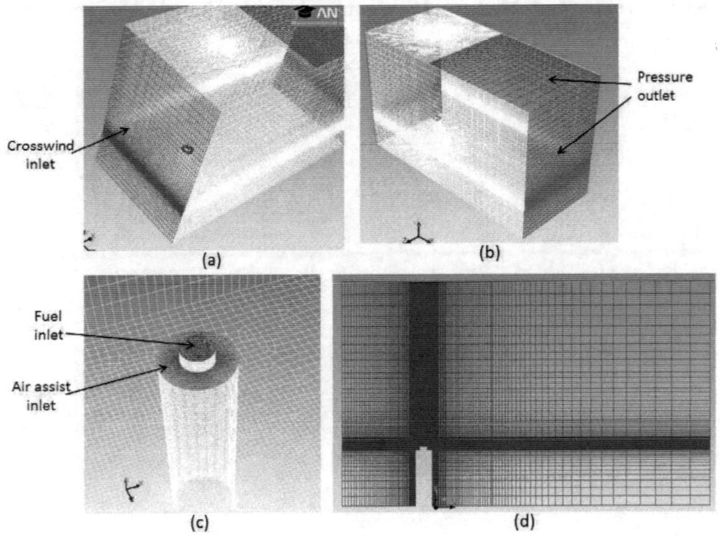

Orthogonal quality of =0.76
grid size = 0.8 million

15AIChE Spring Meeting, Austin, TX

Comparison: EDC and PDF Turbulence-Chemistry Models

EDC (Eddy Dissipation Concept)	PDF (Probability Density Function)
Reactions taking place in the flame are governed by the Arrhenius rates	Reactions are governed by a conserved scalar quantity known as mixture fraction
Incorporates detailed chemical mechanisms	Not so rigorous
Molar concentrations are derived from Reaction rates, which are calculated using ISAT algorithm	Molar concentrations are derived from the predicted mixture fraction fields
Any number of inlet streams can be defined	Only two inlet streams are allowed i.e. Fuel and Oxidizer
Computationally very expensive; requires 5-6 days for convergence	Requires less time for convergence; only 2-3 days

15AIChE Spring Meeting, Austin, TX

Input Conditions for Simulated Air-Assist Cases

Case no.	Total (kg/s)	Temp (°F)	Actual Vent gas LHV (Btu/scf)	Exit velocity (m/s)	Air flow rate (kg/s)	Crosswind (m/s)
A1.1	0.1158	87	2108	0.43	18.8	5.68
A2.1	0.0446	94	2120	0.15	10.54	5.81
A3.1	0.0883	89.3	339	0.58	2.3	5.99
A4.1	0.1181	68.6	584	0.58	6.52	4.06
A5.1	0.0448	56.4	344	0.24	1.1	1.2
A6.1	0.0442	85.9	584	0.21	1.43	7.1

15 AIChE
Spring Meeting, Austin, TX

CFD Modeling: PDF Results with LU 3.0.1 Combustion Efficiency (2010 Tulsa Tests)

Case no.	CE Experiment (%)	CE Simulation (%)	% error
A1.1	96.9	96.93	0.03
A2.1	95.9	95.09	0.84
A3.1	98.3	98.88	0.59
A4.1	97.1	99.23	2.19
A5.1	95.9	99.37	3.62
A6.1	99.4	92.9	6.54
Avg.			2.3

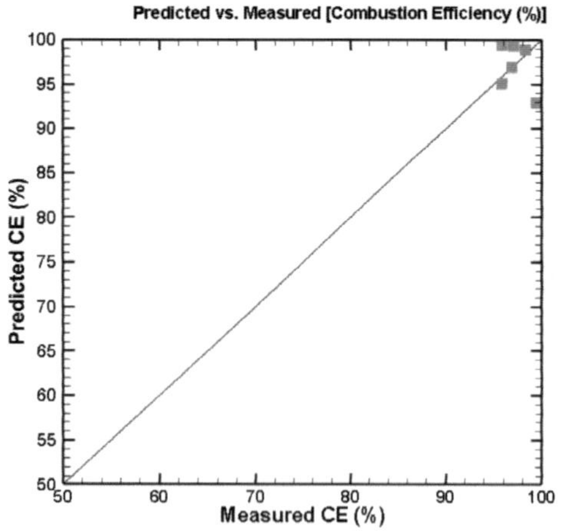

CE predictions: max. 6%, with an avg. of 2%

15 AIChE
Spring Meeting, Austin, TX

CFD Simulation: PDF Results with LU 3.0.1 Soot (2010 Tulsa Tests)

Case no.	BC Experiment (lb/MMBTU)	log BC Experiment (lb/MMBTU)	BC simulation (lb/MMBTU)	log BC Simulation (lb/MMBTU)	Error, a factor of
A1.1	3.05	0.48	4.22	0.63	1.38
A2.1	4.6	0.66	6.19	0.79	1.35
A3.1	3.95	0.60	1.86	0.27	2.12
A4.1	2.68	0.43	1.22	0.09	2.2
A5.1	3.45	0.54	1.55	0.19	2.23
A6.1	8.05	0.91	11.46	1.06	1.42
Avg.					1.78

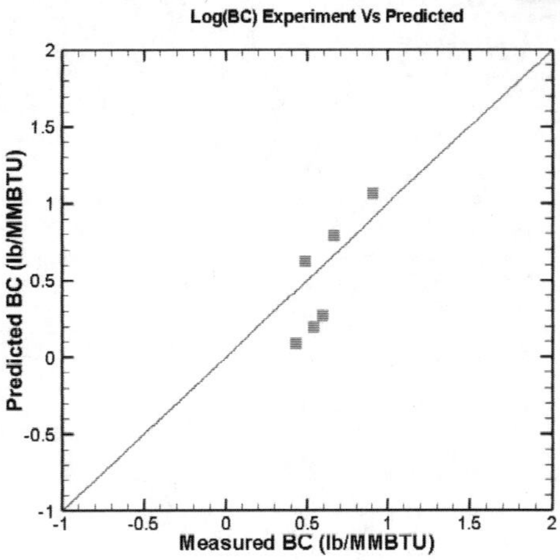

Soot yield predictions: in the right order of magnitude, with in a factor of 2
But PDF predicts virtually no VOCs (emissions are C and CO)

15 AIChE
Spring Meeting, Austin, TX

Geometry For EDC Modeling

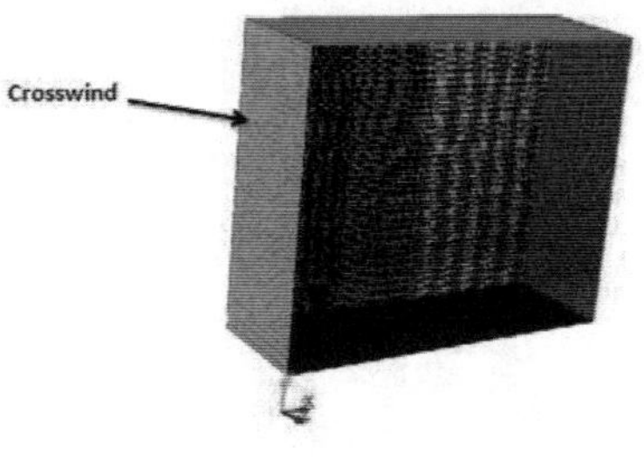

CFD Domain for air-assisted flares

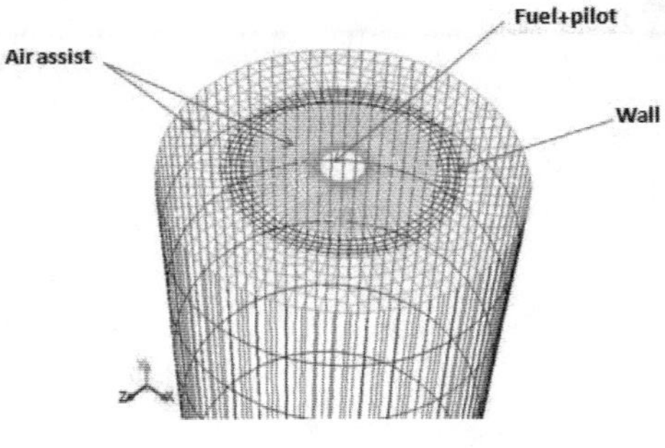

Side view of air-assisted stack showing the inlet surfaces.

15 AIChE
Spring Meeting, Austin, TX

EDC (Eddy Dissipation Concept) Model Results: With LU 3.0.1

Case no.	Propylene (kg/s)	TNG (kg/s)	Nitrogen (kg/s)	Total (kg/s)	Temp (°F)	Actual Vent gas LHV (Btu/scf)	Exit velocity (m/s)	Air flow rate (kg/s)	Crosswind (m/s)
A1.1	0.1158	0.0000	0.0000	0.1158	87	2108.00	0.43	18.80	5.68
A2.1	0.0446	0.0000	0.0000	0.0446	94	2120.00	0.15	10.54	5.81
A4.3	0.0378	0.0039	0.0746	0.1168	84	564.00	0.58	8.5176	4.79

Case no.	CE Experiment (%)	CE Simulation LU 3.0.1 (%)	% Error
A1.1	96.9	97.86	0.99
A2.1	95.9	97.92	2.10
A4.3	94.8	89.81	7.50
Avg.			2.78

Case no.	Soot Experiment (lb/MMBTU)	Soot simulation LU3.0.1 (lb/MMBTU)	Error, a factor of
A1.1	3.05	5.84	1.38
A2.1	4.6	2.24	2.05
A4.3	1.04	0.89	1.16
Avg			**1.53**

CE predictions: comparable, but may lead to convergence difficulties at times
Soot yield predictions: in the right order of magnitude, also within a factor of 2

15 AIChE
Spring Meeting, Austin, TX

VOC Emissions Comparison: EDC with LU 3.0.1

Case	CH4			C3H6			C2H2			C2H4			HCHO			CH3CHO		
#	EXP	SIM	factor of	EXP	SIM	factor of	EXP	SIM	factor of	EXP	SIM	factor of	EXP	SIM	factor of	EXP	SIM	factor of
A1.1	2.99E-04	3.29E-03	11.01	2.27E-03	6.71E-03	2.96	1.42E-04	1.28E-03	9.01	6.97E-05	1.31E-03	18.84	1.05E-04	8.24E-04	7.84	6.98E-05	2.55E-04	3.65
A2.1	1.26E-04	6.71E-06	18.77	9.58E-04	4.33E-04	2.21	4.38E-05	2.68E-04	6.12	1.98E-05	6.79E-03	343.07	2.90E-05	2.89E-03	99.78	1.84E-05	7.35E-05	4.00
A4.3	1.97E-04	3.82E-04	1.94	1.49E-03	1.95E-03	1.31	6.79E-05	1.81E-03	26.60	3.07E-05	3.02E-03	98.33	4.51E-05	5.30E-03	117.53	2.85E-05	2.39E-04	0.12
Avg error			10.57			2.16			13.91			153.42			75.05			2.59

EDC model with LU 3.0.1 tends to over-predict VOCs

15 AIChE
Spring Meeting, Austin, TX

Geometry used for Modeling Air-assisted flares in *2012-13*

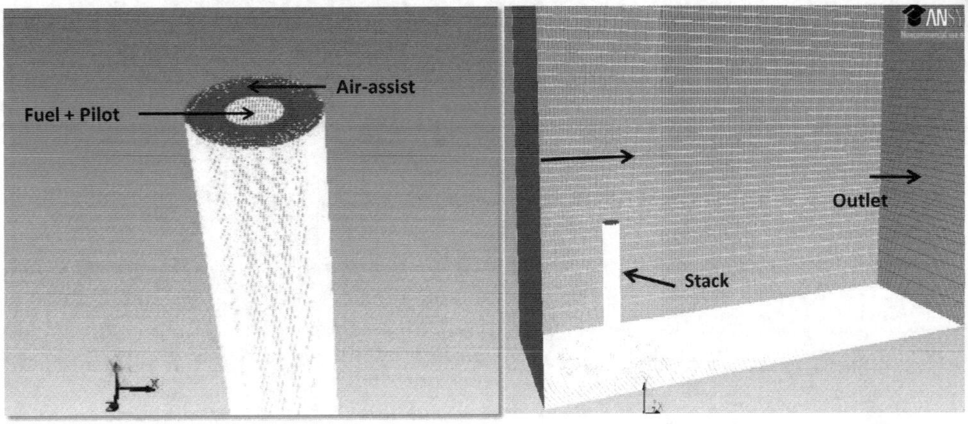

(Concentrated, localized fuel leads to a sustained flame with EDC)

15 AIChE
Spring Meeting, Austin, TX

Comparison of measured and CFD simulation results Destruction & Removal Efficiency (DRE, %)

Case #	TCEQ Measurement	AQRP 10-022 Report		2013 Simulation	
		DRE (%)	Error(%)	DRE (%)	Error(%)
A1.1	98.05%	78.65%	-19.79%	99.80%	1.78%
A2.1	97.15%	84.14%	-13.39%	99.38%	2.30%
A2.3	96.19%	87.42%	-9.12%	99.54%	3.48%
A2.4	92.26%	81.82%	-11.32%	97.97%	6.19%
A2.5	95.09%	82.95%	-12.77%	97.61%	2.65%
A3.3	88.13%	Flame not sustained in CFD	-	91.73%	4.08%
A3.6	91.73%		-	99.31%	8.26%
A4.3	93.77%		-	94.20%	0.46%
A6.4	90.82%		-	96.65%	6.42%
A6.5	88.29%		-	97.40%	10.32%
Average Error (absolute)		13.28%		4.59%	

*Test cases with A/F mass ratio<28

15 AIChE
Spring Meeting, Austin, TX

Comparison of measured and CFD simulation results Combustion Efficiency (CE, %): EDC With LU 1.1

Case #	TCEQ Measurement	AQRP 10-022 Report		2013 Simulation	
		CE (%)	Error(%)	CE (%)	Error(%)
A1.1	96.72%	59.98%	-37.99%	96.40%	-0.33%
A2.1	95.54%	63.97%	-33.04%	96.91%	1.43%
A2.3	94.02%	77.18%	-17.91%	94.06%	0.04%
A2.4	87.64%	72.42%	-17.37%	92.36%	5.39%
A2.5	91.84%	52.54%	-42.79%	93.03%	1.30%
A3.3	84.68%	Flame not sustained in CFD	-	81.48%	-3.78%
A3.6	88.88%		-	92.61%	4.20%
A4.3	91.50%		-	87.32%	-4.57%
A6.4	86.32%		-	84.16%	-2.50%
A6.5	81.55%		-	83.86%	2.83%
Average Error (absolute)		**29.82%**		**2.64%**	

*Test cases with A/F mass ratio<28

CE predictions: max. 5%, with an avg. of 3%
Can simulate lower LHV flares

15 AIChE
Spring Meeting, Austin, TX

Non-Assisted Lab-Scale Flare Experiments at Energy & Emissions Research Lab, Carleton University, Ottawa, Canada

Photo showing sampling hood and flare at the Energy & Emissions Research Lab. Flare Facility at Carleton University

Fuel: C_1-C_4 Alkane Mixtures & Ethylene

Species	Purity (%)	Average 6-Mix (%)	Heavy 4-Mix (%)	Ethylene (%)
Methane	99.0	85.24	74.54	-
Ethane	99.0	7.06	15.47	-
Propane	99.0	3.11	6.83	-
Butane	99.0	1.44	3.16	-
Carbon Dioxide	99.0	1.91	-	-
Nitrogen	99.999	1.24	-	-
Ethylene	99.5	-	-	100

A turbulent diffusion non-premixed flame burning hydrocarbon mixtures relevant to low-momentum flares ubiquitous in the oil and gas industry

15 AIChE
Spring Meeting, Austin, TX

Moss-Brookes Soot Model & Model Parameters Tuning

$$\frac{dM}{dt} = \underbrace{M_P C_\alpha \left(\frac{X_{prec}P}{RT}\right)^l \exp\left\{-\frac{T_\alpha}{T}\right\}}_{\text{Nucleation}}$$

$$\underbrace{+ C_\gamma \left(\frac{X_{sgs}P}{RT}\right)^m \exp\left\{-\frac{T_\gamma}{T}\right\}\left[(\pi N)^{1/3}\left(\frac{6M}{\rho_{soot}}\right)^{2/3}\right]^n}_{\text{Surface Growth}}$$

$$\underbrace{- C_{oxid} C_\omega \eta_{coll}\left(\frac{X_{CH}P}{RT}\right)\sqrt{T}(\pi N)^{1/3}\left(\frac{6M}{\rho_{soot}}\right)^{2/3}}_{\text{Oxidation}}$$

$$\frac{\partial}{\partial t}(\rho Y_{soot}) + \nabla\cdot(\rho \vec{v} Y_{soot}) = \nabla\cdot\left(\frac{\mu_t}{\sigma_{soot}}\nabla Y_{soot}\right) + \frac{dM}{dt}$$

$$\frac{\partial}{\partial t}(\rho b^*_{nuc}) + \nabla\cdot(\rho \vec{v} b^*_{nuc}) = \nabla\cdot\left(\frac{\mu_t}{\sigma_{nuc}}\nabla b^*_{nuc}\right) + \frac{1}{N_{norm}}\frac{dN}{dt}$$

$$\frac{dN}{dt} = \underbrace{C_\alpha N_A \left(\frac{X_{prec}P}{RT}\right)^l \exp\left\{-\frac{T_\alpha}{T}\right\}}_{\text{Nucleation}} \underbrace{- C_\beta \left(\frac{24RT}{\rho_{soot} N_A}\right)^{1/2} d_p^{1/2} N^2}_{\text{Coagulation}}$$

The set of constants proposed by Brookes and Moss for methane flames are given below:

- $C_\alpha = 54\ s^{-1}$ (model constant for soot inception rate)
- $T_\alpha = 21000$ K (activation temperature of soot inception)
- $C_\beta = 1.0$ (model constant for coagulation rate)
- $C_\gamma = 11700\ \mathrm{kg \cdot m \cdot kmol^{-1} \cdot s^{-1}}$. (surface growth rate scaling factor)
- $T_\gamma = 12100$ K (activation temperature of surface growth rate)

Y_{soot} = soot mass fraction
M = soot mass concentration (kg/m-3-)
b^*_{nuc} = normalized radical nuclei concentration
N = soot particle number density (particles/m^3)
$N_{norm} = 10^{15}$ particles

15 AIChE Spring Meeting, Austin, TX

ANSYS FLUENT Theory Guide Release 14.0 ; November 2011. ANSYS, Inc.

CFD Simulation: PDF Results
Carleton University 2014 Experimental Data

Soot Results with Reynolds Number Correction Equation For Moss-Brookes Soot Model & Lee Soot Oxidation Model

- Moss-Brookes soot model mainly rely on two model constants, constant for soot inception rate (C_α) and constant for soot coagulation rate (C_β)
- These model constants determined **through numerical modeling of a laminar flame** for which experimental data exists (Fluent Theory Guide)
- As these constant are determined from laminar flame experimental data, they may not hold for high velocity turbulent jet flame

Reynolds Number Correction Equation For Moss-Brookes Soot Model

$$C_\alpha = 377.68 - 39.44\, ln(Re)$$

$$C_\beta = 54/C_\alpha$$

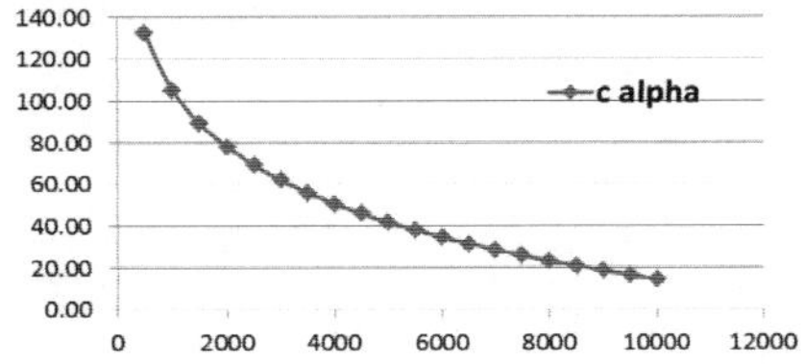

15 AIChE Spring Meeting, Austin, TX

CFD Simulation: PDF Results
Carleton University 2014 Experimental Data

Soot Yield with/without Reynolds Number Correction Equation

Without Correction | **With Correction**

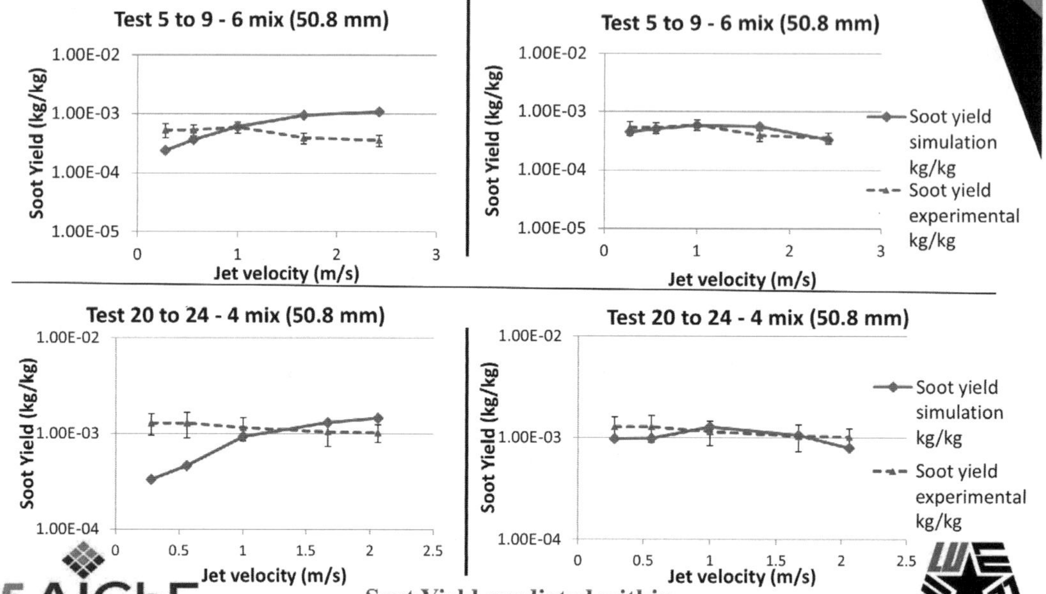

Soot Yield predicted within uncertainty limits after correction equation

15 AIChE Spring Meeting, Austin, TX

CE & DRE Results for CU Data-EDC Model

CE&DRE for 6-Mix. Fuel(50.8mm)Jet Diameter

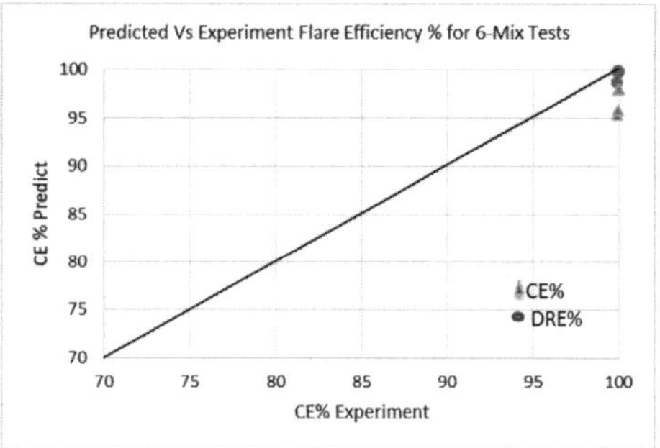

CE&DRE for 4-Mix. Fuel(50.8mm)Jet Diameter

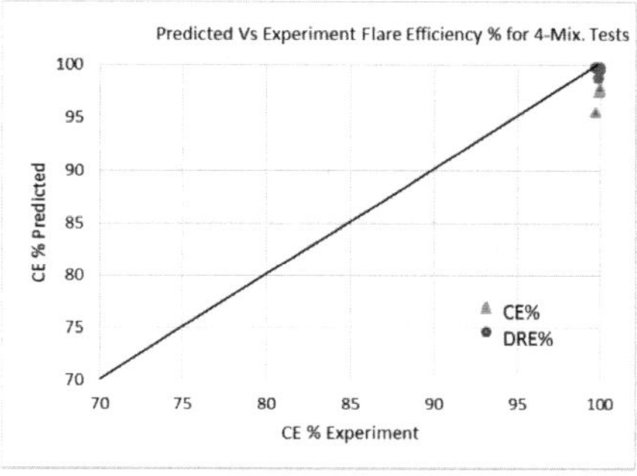

CE% for Pure Ethylene Fuel(50.8mm) Jet Dia.

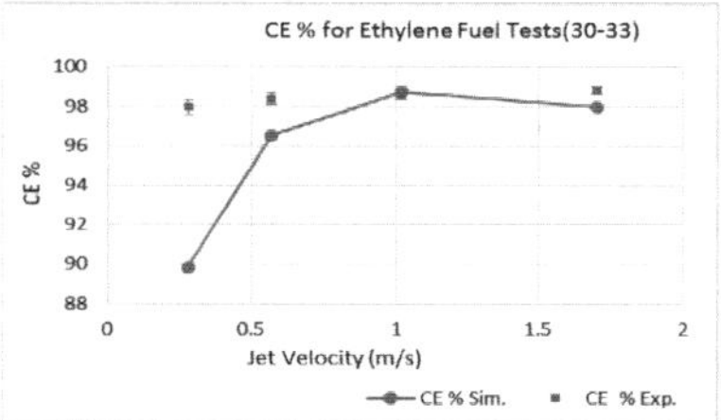

Generally in good agreement but tends to under-predict both DRE & CE

15 AIChE Spring Meeting, Austin, TX

EDC Results- Carleton University 2014 Tests

CO2 Yield Results for 3 Different Fuel (50.8 mm Jet-Diameter)

AVG 6-Mix. Fuel (Tests6-9) HVY 4-Mix. Fuel(20-24) Pure Ethylene Fuel(30-33)

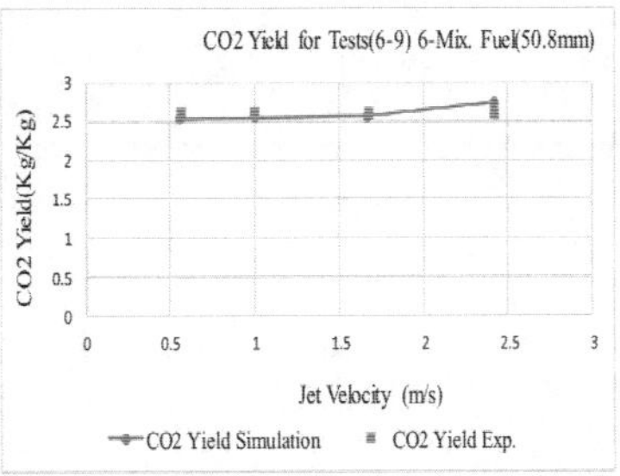

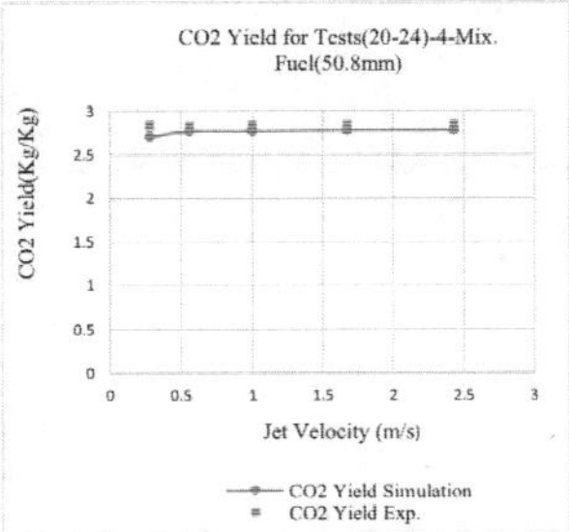

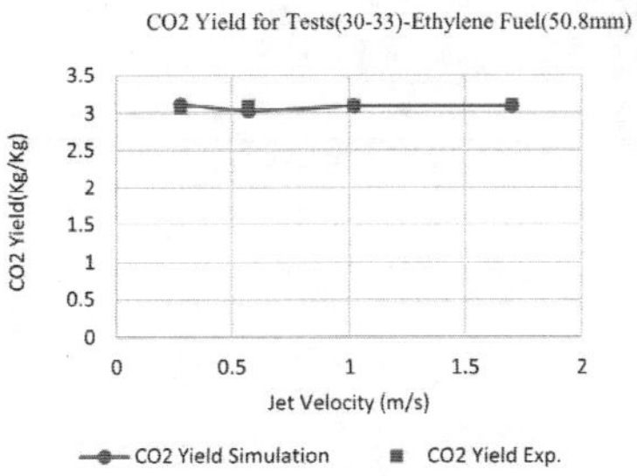

CO2 Yield increases with jet velocity

15 AIChE Spring Meeting, Austin, TX

CO and CH4 Yield Results for HVY 4-Mix. Fuel (50.8 mm Jet-Diameter)

CO Yield

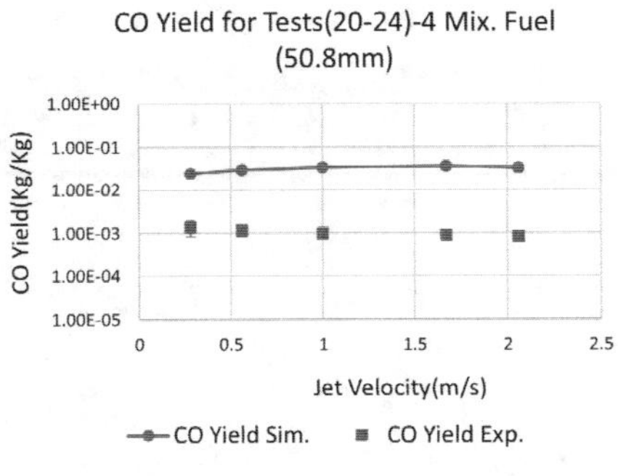

CH4 Yield

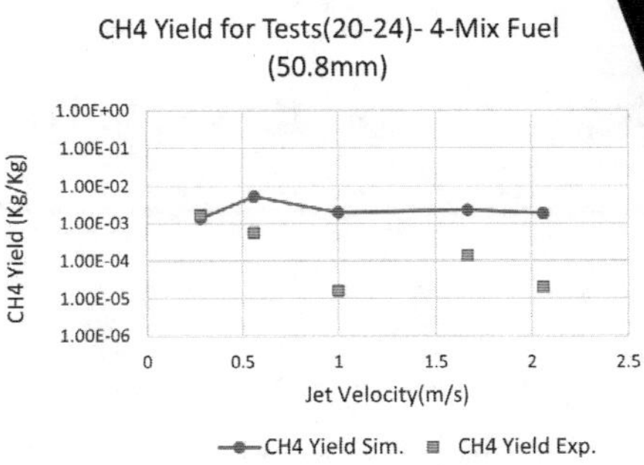

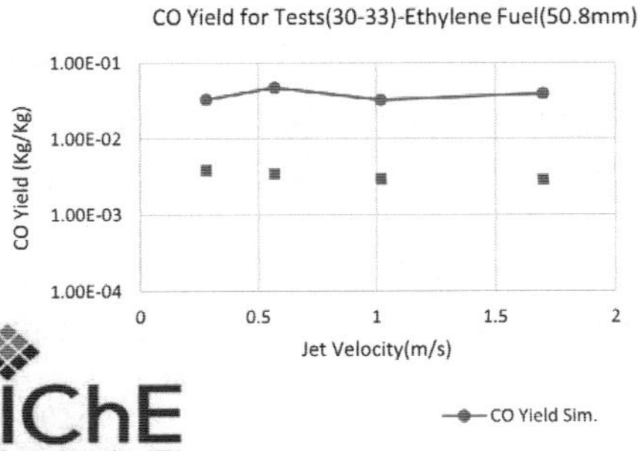

CO & CH4 Yields are over-predicted by the EDC model.

15 AIChE Spring Meeting, Austin, TX

Soot Yield Results for 3 Different Fuel (50.8 mm Jet-Diameter)

AVG 6-Mix. Fuel (Tests 6-9) | HVY 4-Mix. Fuel(20-24) | Pure Ethylene Fuel(30-33)

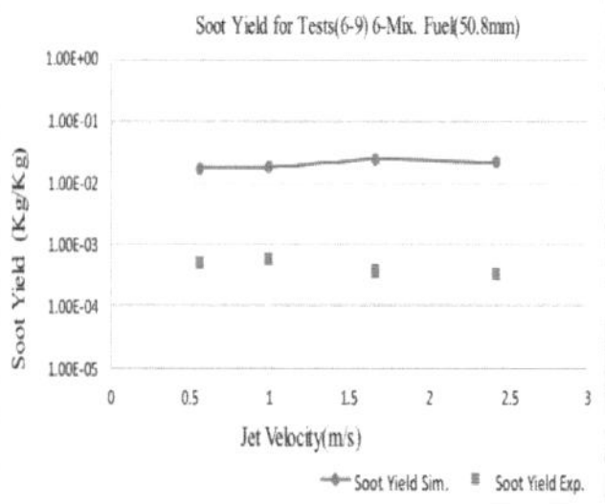

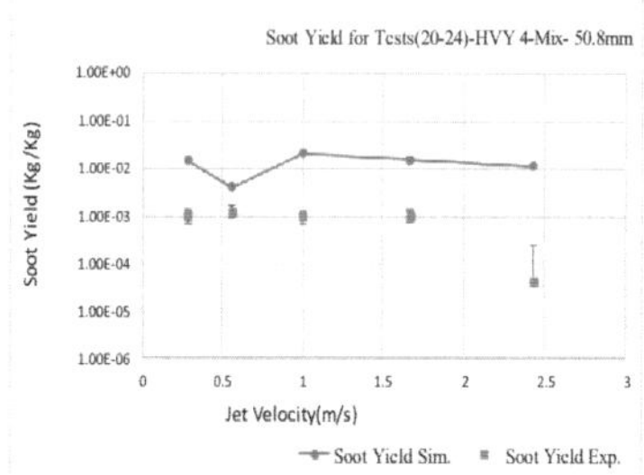

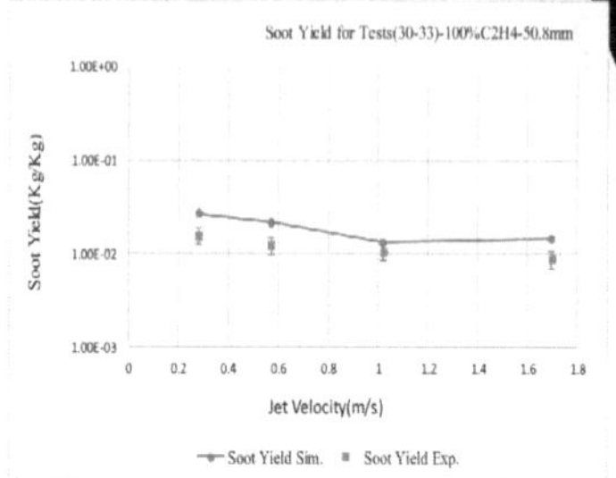

Soot Yield based on the EDC Turbulence Chemistry model & Fennimore-Jones Soot Oxidation Model :

- over-predicted for methane-based flares
- in agreement with experiments for ethylene flares

15 AIChE Spring Meeting, Austin, TX

Observations and conclusion

PDF Model	EDC Model
CE predictions: max. 6%, with an avg. of 2% DRE is always over-predicted	Under-predicts both DRE & CE but in good agreement
Easy to converge	Convergence difficulties at times
Virtually predicts no VOCs	Reasonably predicts VOCs depending on the mechanism used
Soot Yield: ❖ Predicted within uncertainty limits after correction equation for CU experimental data ❖ Predicted in the right order of magnitude, with in a factor of 2 for TCEQ test cases	Soot yield: ❖ For CU data, ○ Over-predicted for methane-based flares ○ In agreement with experiments for ethylene flares ❖ Predicted in the right order of magnitude, within a factor of 2 for TCEQ test cases
Temperature ranges from [1760-1820]K methane-based flares and [1935-1990] K for ethylene flares	Temperature ranges from [2181-2226] K for methane-based flares range and [2378-2395]K for ethylene flares.
Lee Soot Oxidation Model works better.	Fennimore-Jones Soot Oxidation Model works better.

LU3.0.1	LU 1.1
Predicts VOCs and soot	Predicts VOCs and NOx
Handles C1-C4 Species	Handles C1-C3 species
EDC model tends to over-predict VOCs	EDC model tends to predict reasonable VOCs
Convergence difficulty in case of lower LHV flares	Can simulate lower LHV flares
Based on USC II mechanism	Based on GRI 3.0 and USC I

15 AIChE Spring Meeting, Austin, TX

Lamar University CFD Lab

- Graduate Students
  - Vijaya Damodara, Kader Rasel, Raj Alphones, Lan Liu (PhD), Ajit Patki, (DE), Hashim Almaryani, Anan Wang (MES)
- Cutting Edge High Performance Computing (HPC) Cluster
  - 3 Thinkmate HPX XS8-2460 servers, Intel C602 Chipset - Dual 1-Gigabit Ethernet - 4U/Tower GPU, 2 x Six-Core Intel Xeon Processor E5-2630 v2 2.60GHz 15MB Cache (80W), 8 x 8GB PC3-12800 1600Mhz DDR3 ECC Registered DIMM
  - 3 X 12 core Dell servers; Intel Xeon X5670 @2.93GHz, More than 50 high speed processors, Up to 10GBs/second of data transfer speed for faster parallel computing
  - 3x24 Core Dell servers; Intel Xeon E5-2630V2 @2.6 GHz

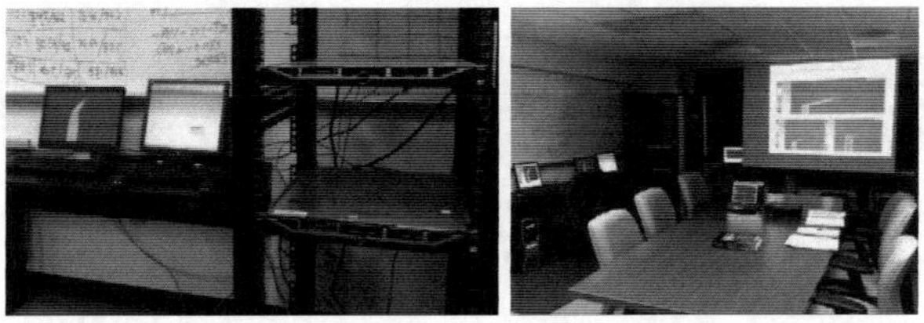

15 AIChE
Spring Meeting, Austin, TX

References

- McDaniel, M., Flare Efficiency study, EPA-600/2-83-052, July 1983.
- Pohl, J., R. Payne, and J. Lee. 1984. Evaluation of the efficiency of industrial flares: Test Results. EPA-600/2-84-095. Prepared for U.S. EPA Office of Research and Development by Energy and Environmental Research Corporation (May).
- EPA. (2014). Method 9-Visual Determination of the Opacity Of Emissions From Stationary Sources, "Environmental Protection Agency, 40 CFR Ch. I (7–1–09 Edition), Pt. 60, App. A–4, Meth.9. http://www.epa.gov/ttn/emc/methods/method9.html.
- EPA. (1993).Visible Emissions Field Manual, EPA Methods 9 and 22, US. ENVIRONMENTAL PROTECTION AGENCY, Stationary Source Compliance Division, Office of Air Quality Planning and Standards, Washington, DC 20460.
- D. Allen and V. Torres, 2010 TCEQ Flare Study Final Report & Appendices," The University of Texas at Austin, TCEQ PGA No. 582-8-86245-FY09-04 and Task Order No. UTA10-000924-LOAT-RP9, Aug. 1, 2011.
- D. Corbin & M. Johnson. "Summary of Flare Efficiency and Soot Emission Rate Measurement Results at Carleton University Energy & Emissions Research Lab", TCEQ GAD Project number 582-10-94307-FY14-06, PI, D. Chen, Aug. 2014.
- K. Singh, P. Gangadharan, D. Chen, H. Lou, X. Li & P. Richmond (2014), "Computational fluid dynamics modeling of laboratory flames and an industrial flare", Journal of the Air & Waste Management Association, 64:11, 1328-1340.
- K. Singh, T. Dabade, H. Vaid, P. Gangadharan, D. Chen, H. Lou, X. Li, K. Li, C. Martin, "Computational Fluid Dynamics Modeling of Industrial Flares Operated in Stand-By Mode," Industrial Flares special issue, Ind. & Eng. Chem. Research, 51 (39), 12611-12620, , October, 2012.

15 AIChE
Spring Meeting, Austin, TX

Acknowledgements

Financial support from TCEQ Grant for Activities Program (Project # 582-10-94307-FY14-06), TCEQ Supplemental Environmental Program (SEP Agreement no. 2009-009), and the Texas Air Research Center (TARC Grant #079LUB0096A) are gratefully acknowledged.

Special thanks are due to Ed Fortner and Scott Herndon of Aerodyne Research, Inc. (ARI) for providing numeric soot data for 2010 JZ flare campaign.

Thank you!

15 AIChE
Spring Meeting, Austin, TX

Removal of Sulfates through Precipitation

William E Moore
Westech engineering

15 AIChE
Spring Meeting, Austin, TX

WesTech
Process Equipment. Process Driven.

Removal of Sulfates through Precipitation

- Sulfates are naturally occurring
- At low levels not hazardous to humans or animals
- However "too much of a good thing" causes taste issues and diarrhea

15 AIChE
Spring Meeting, Austin, TX

WesTech
Process Equipment. Process Driven.

Sources

- Leaching sulfur out of soil
- Mines – Almost all ores are metal sulfides. The metal is extracted and the sulfide oxidizes to sulfate.
- Oil and natural gas - Sour oil or natural gas contains sulfide. The sulfide is very corrosive so it is removed from the organics and oxidized to sulfate.

15 AIChE Spring Meeting, Austin, TX

WesTech Process Equipment. Process Driven.

Standards

- World Health Organization (WHO) are <500 ppm
- EPA guidelines are the same but many states are requiring <250 ppm for potable (drinking) water

15 AIChE Spring Meeting, Austin, TX

WesTech Process Equipment. Process Driven.

Sulfate Removal

- Most treatments for sulfate removal (ion exchange, RO, NF, evaporation, etc.) transfer the sulfate from a dilute solution to a concentrated solution. This reduces volume but does not solve the problem.
- Precipitation can remove most of the sulfates in the form of a non-hazardous solid so that all liquid can be discharged.

15 AIChE Spring Meeting, Austin, TX

WesTech Process Equipment. Process Driven.

Sulfate Precipitation

- Unfortunately sulfates are fairly soluble so removal may not be straight forward.
- A typical process would be :

Lime softening

Sulfate concentration

Calcium Sulfate Precipitation

Aluminum addition

15 AIChE Spring Meeting, Austin, TX

WesTech Process Equipment. Process Driven.

Lime Softening

- "Hard water" is water with a high concentration of calcium, magnesium and bicarbonates
- If you add lime (Ca(OH)2) to water with Ca(HCO)3 present most of the calcium will precipitate as CaCO3! This reduces the hardness so the water is less prone to scaling.

15 AIChE Spring Meeting, Austin, TX

WESTECH Process Equipment. Process Driven.

Lime Softening

- The $Mg(HCO_3)_2$ will also react with the lime to form MgCO_3 which will precipitate.
- If the sulfate concentration is above 2800 ppm CaSO_4 will also precipitate
- Lime softening can take place in an open tank but is usually done in a clarifier.

15 AIChE Spring Meeting, Austin, TX

WESTECH Process Equipment. Process Driven.

Lime Softener

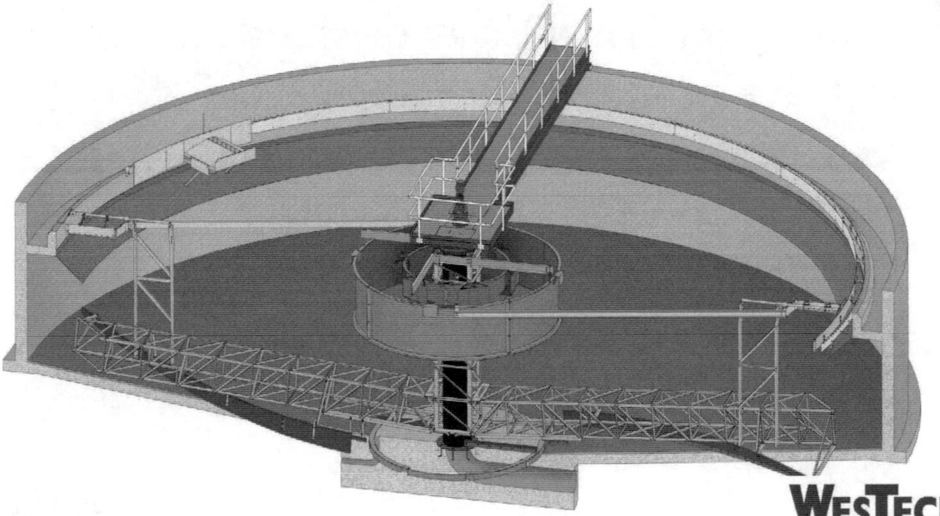

WesTech
Process Equipment. Process Driven.

15 AIChE
Spring Meeting, Austin, TX

Lime Softening

- In addition to reducing Mg and Ca concentrations, lime softening will also precipitate many heavy metals such as Fe, Cu, Ni, Zn, etc.

WesTech
Process Equipment. Process Driven.

15 AIChE
Spring Meeting, Austin, TX

Sulfate Concentration

- In another of chemistry's mysteries, if you can increase sulfate concentration above ~ 3500 ppm (supersaturation) and then precipitate with lime you can drive the sulfate concentration to 1500 ppm.
- An economic way to increase sulfate concentration is through an UF membrane.

15 AIChE
Spring Meeting, Austin, TX

WesTech
Process Equipment. Process Driven.

NF

- The sulfate concentration in the reject is controlled by the split between permeate and reject flows. If the influent sulfate concentration is 1400 and the reject flow is half that of the permeate then the sulfate concentration goes to 4200.
- Supersaturated fluids are tricky so usually the max sulfate concentration is 4200-4500 ppm.

15 AIChE
Spring Meeting, Austin, TX

WesTech
Process Equipment. Process Driven.

NF

- After some study it was found that a nano-filter is a good way to concentrate sulfates.
- A nano-filter is a polymeric membrane with tightly controlled pore sizes. Essentially it splits the influent into two streams. It will retain 98% of the calcium and sulfate in the reject while allowing 99% of the sodium to pass through into the permeate.

WESTECH
Process Equipment. Process Driven.

15 AIChE
Spring Meeting, Austin, TX

Membrane Systems

Reverse Osmosis Equipment

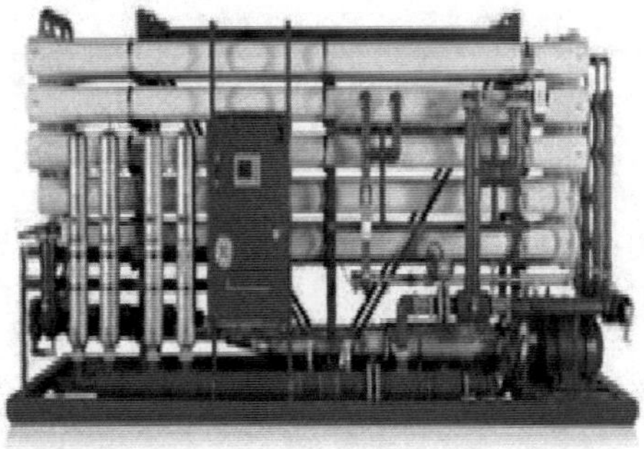

Remove dissolved impurities from water

WESTECH
Process Equipment. Process Driven.

15 AIChE
Spring Meeting, Austin, TX

Desupersaturation

- Now that the sulfates are supersaturated, you add lime to raise the pH to ~12.0. The additional calcium reacts with the sulfate and CaSO_4 (gypsum) precipitates.
- The high pH allows the reaction to continue past the solubility limits. Concentration of 1500 -1800 ppm are easily achieved.

15 AIChE
Spring Meeting, Austin, TX

WesTech
Process Equipment. Process Driven.

Desupersaturation Equipment

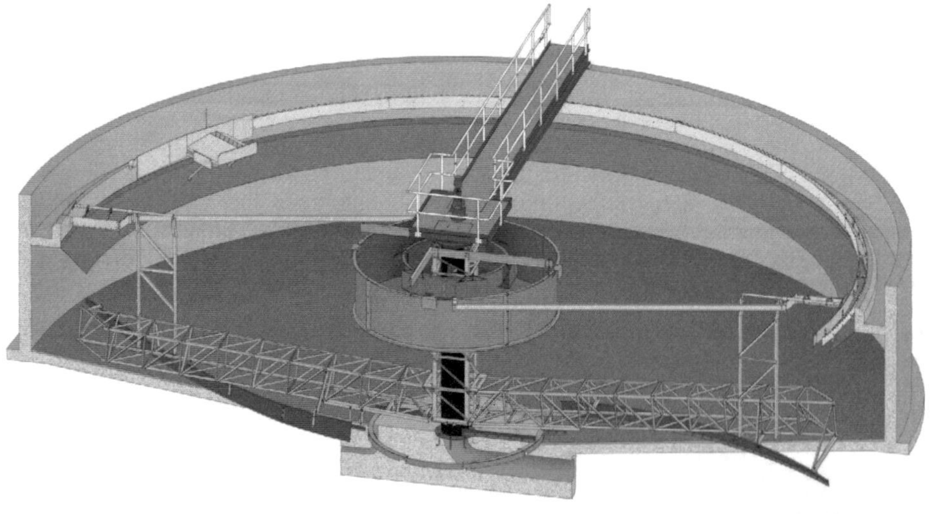

15 AIChE
Spring Meeting, Austin, TX

WesTech
Process Equipment. Process Driven.

Further Sulfate Reduction

- If you dilute the desupersaturation effluent with UF permeate you will obtain a sulfate concentration of 500-600 ppm. If this will meet the permit you are done.
- If it does not meet permit levels then we have to do one more precipitation with aluminum.

15 AIChE Spring Meeting, Austin, TX

WesTech Process Equipment. Process Driven.

Aluminum Precipitation

- To accomplish this reaction lime and aluminum are added to another clarifier
- This reaction can drive sulfate concentrations below 100 ppm but the aluminum is expensive so usually you drive the reaction to the maximum allowed by the permit.

15 AIChE Spring Meeting, Austin, TX

WesTech Process Equipment. Process Driven.

Sulfate Reduction Through Aluminum Precipitation

- At high pH aluminum will react with CaSO_4 to form a mineral called Ettringite (3CaO·3CaSO_4·Al_2O_3·31H_2O)
- Many different forms of Al have been used (alum, sodium aluminate, cement additives, even aluminum balls). The form of the aluminum does not seem to matter so whatever is cheaper is used.

15 AIChE Spring Meeting, Austin, TX

WESTECH Process Equipment. Process Driven.

System Drawbacks

- Lots of solids handling. Tons/day of solids can be generated by a fairly small stream.
- The solids get progressively harder to handle as you go through the system.
- Very high cost.

15 AIChE Spring Meeting, Austin, TX

WESTECH Process Equipment. Process Driven.

System Benefit

- 98% of the water is recovered
- The water will meet all regulatory limits. You can convert a hazardous waste into drinking water.
- There are no waste fluids. The only wastes are non-hazardous solids.

15 AIChE
Spring Meeting, Austin, TX

WesTech
Process Equipment. Process Driven.

MITIGATING VARIABLE TDS LEVELS IN WASTEWATER BIOTREATMENT

By William Celenza, Melissa Owens and Janani Sankar

KBR Inc,

Houston, TX

Introduction

Wastewaters characterized by high total dissolved solids (TDS) levels and elevated organic content are produced by facilities in the chemical industry, especially the chloralkali sector, agro-food industries, petroleum, natural gas fracturing, and textile / leather industries. Even the occasional intrusion of seawater such as from use as a backup supply for fire water has resulted in elevated TDS levels. Table 1 shows TDS levels in various process streams and Table 2 lists typical inorganic constituents of TDS.

Although inorganic constituents comprise most of the total, dissolved solids can originate from organic sources such as runoff, silt, plankton, and sewage. High TDS effluents have been traditionally treated by physical and chemical means, since one of the most inhibitory substances in biotreatment is salinity. However following pretreatment, water characterized by elevated organic content combined with high TDS effluents can be treated by biological means to achieve carbon, nitrogen and phosphorous removal; with even concentrations of over 100,000+ ppm of TDS found to permit biodegradation (Diaz, et al). Although in general, the higher the TDS level the lower the biological stability to environmental change such as hydraulic shocks (Pophali et al). For comparison in a recent state of Pennsylvania regulatory program, a "high TDS" wastewater discharge is considered a TDS concentration that exceeds 2,000 mg/l. Also, Royal Commission Environmental Regulations limit TDS in effluent discharged to Municipal Authorities to 2,000 mg/l, while limiting treated effluent discharges to coastal waters to no more than 1,400 mg/l of salinity above ambient in the Red Sea or Arabian Gulf. Stable performance of primary and biological treatment steps is essential for meeting effluent discharge limits and for optimizing performance of post treatment steps. Some states such as New Jersey, Pennsylvania, and Ohio apply effluent standards to wastewater discharges as low as 500 mg/l TDS to protect the quality of public drinking water supply sources. Downstream treatment such as reverse osmosis, ultra filtration, or evaporation/crystallization may need to be considered to meet such stringent criteria. Realizing the TDS level in raw waters varies with origin, Figure 1 indicates typical ranges for classification of water quality according to total TDS concentration; TDS ranges for classifications presented may vary per reference source and by application, i.e. application for irrigation versus drinking water. For example, the TDS levels in a groundwater source may vary from 100 ppm, in the freshwater range, to 50,000 ppm in the brine range (California State Water Resource Control Board).

Table 1 - Example TDS Concentrations in Process Water Streams

Source Water Description	TDS,ppm	Comments
Fracking Water	800 to 300,000	
Bitterns	240,000	From saline water (brine).
HCL Scrubber Effluent	150,000	
High Salt WWTP Effluent	100,000	
Desal RO Reject	70,000	2 Stage RO
Desal Reject (Thermal)	67,500	
Seawater	30,000 to 40,000	
Inorganic WWTP Effluent	10,000+	
Air Pollution Control Scrubbers	50 to 10,000+	Depending on purge rates
Organic WWTP Effluent	2,500	
Sour Water - FCCU	1,200	1,200 ppm H_2S, 250 ppm NH_3
Cooling Tower Blowdown	500 to 2,500	Fresh Water
Sour Water – Vacuum Gas Oil HydroTreating	500	2,000 ppm H_2S, 500 ppm NH_3
Sanitary Wastewater	300	
Boiler Blowdown	100	
Fresh Water Cooling Tower Makeup	75 to 750	50 ppm Ca, 50-300 ppm Cl
Stripped Sour Water	50	10 ppm H_2S, 25 ppm NH_3
Process Use (Desal)	10	Plant Water
Rainwater	<10	
Process Use (Demin)	4	Steam System Make up
Distilled Water	<1.5	

Table 2 – Typical Inorganic Constituents of TDS

Cations	Anions	Metals
Calcium	Bicarbonates	Aluminum
Sodium	Carbonates	Arsenic
	Chlorides	Iron
	Nitrates	Lead
	Phosphates	Magnesium
	Sulfates	Manganese

Figure 1 - Classifications of Water Quality per TDS level (1)

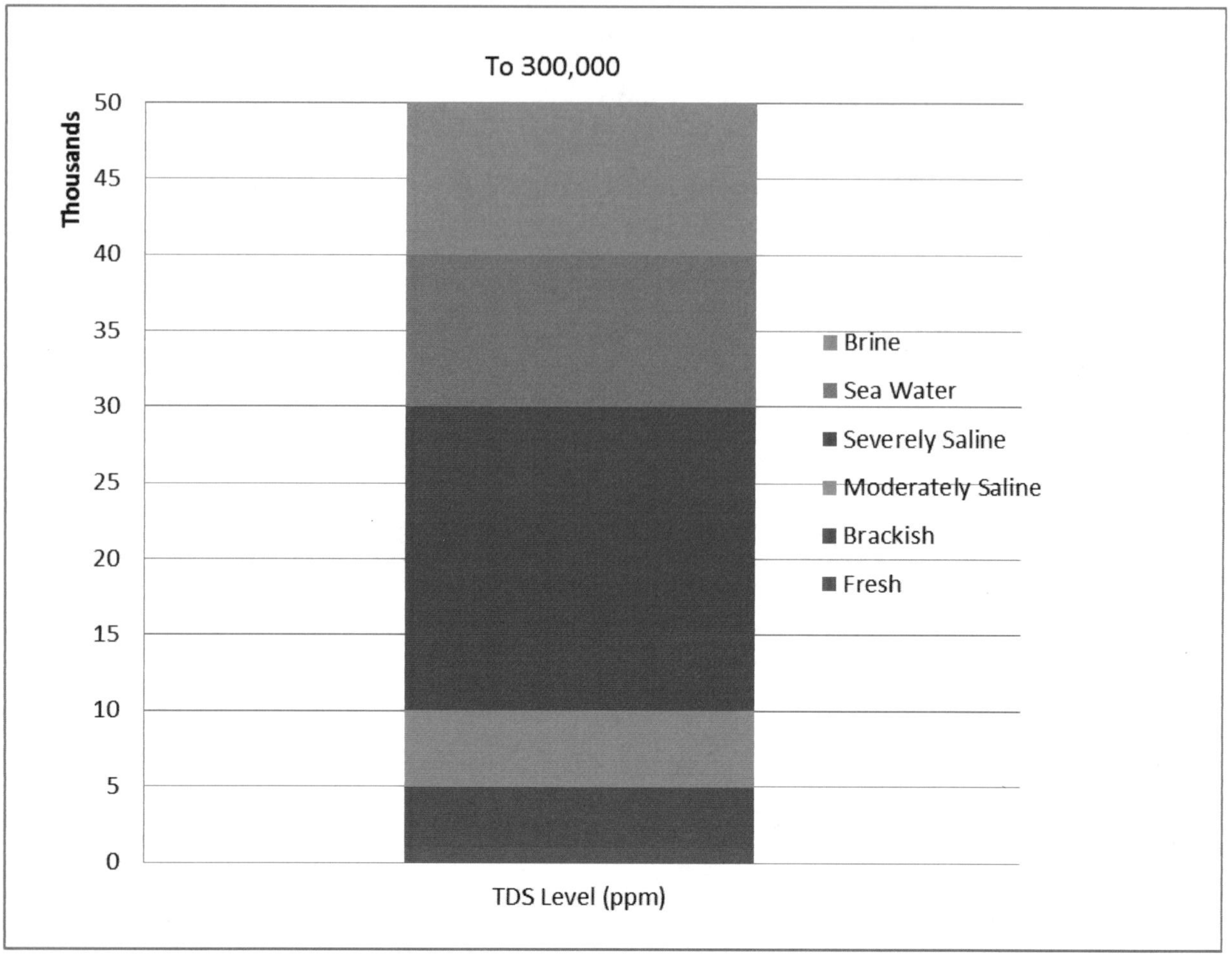

Notes:
(1) (Fair, GM et al.)

For high TDS wastewater, biological performance depends on the proper adaptation of the wastewater's biomass to changes in wastewater conditions. A critical design consideration for TDS treatment is that concentrations are often subject to fluctuations, including variation in the rate of change. When TDS is not maintained within acceptable limits, the operability of biological treatment systems can be adversely impacted.

To effectively treat wastewaters, an acceptable level of influent TDS fluctuation must be defined. The determination of acceptable variability is critical to the stability of a biological treatment system for high strength TDS and organic wastewaters, and thus the maintenance of required operating efficiencies.

This paper discusses the effects of TDS concentration changes when aerobically treating combined high TDS and high organic strength wastewater. Literature sources are surveyed to define the limit to observed biomass response when subject to step changes in influent TDS concentration. Industry design margins are discussed for treatment systems with respect to mitigating variable TDS levels. Finally process design options are discussed, using a wastewater treatment system from a chloralkali by-product manufacturing facility as an example.

Shock Effect on Biomass

The efficiency of a biological treatment process depends on the health of a system's biomass, the total microorganism population. The health of a biomass is monitored directly by measuring certain characteristics, and indirectly by the resultant constituent removal capacity. For instance, some of the main characteristics are listed in Table 3. Also included in Table 3 is the observed impact on biomass characteristics by a sudden change in influent TDS concentration. In general, a TDS shock load is any influent change that can have an inhibitory effect, keeping in mind that reductions in salt concentration can sometimes cause more severe effects on biomass then increases. In particular, a "shock" loading can be defined in terms of a condition of guarantee for a licensor biotreatment unit. For instance, a TDS fluctuation range of 20% per day, either over or under the average influent, has been a required operational constraint to guarantee a COD removal efficiency (Migliavacca). However, the analysis later in this paper also considers shock load in relation to a target TDS level; and the environmentally favored microorganism populations that would dominate near this level.

Several lab scale and pilot scale experiments show the effect of TDS shock loadings on a biological treatment system. For instance, the osmotic pressure of water increases about 8 bar for every 10,000 ppm of salt increase, and thus at 50,000 ppm the osmotic pressure would be above 40 bar (Winson et al). Therefore, cell lysis and the resulting impacts outlined in Table 3 can result. Salinity has a direct impact on oxygen solubility affecting viscosity and air bubble coalescence. The variability of TDS can influence oxygen transfer and impact bio kinetics, as well as reduce the effectiveness of applying air bubbling to alleviate membrane fouling in submerged MBR systems (Winson et al). When TDS increases, the density of the biomass increases. At salt concentrations of 10,000 ppm and 15,000 ppm the density would be 0.7% and 1.1%, respectively, above freshwater increasing at 50,000 ppm to 3.4% (Winson et al). The greater buoyant force exerted by the higher density would result in weaker biological floc structure and hence higher turbidity (Woolard and Irvine). In addition, a greater tendency of scaling and coagulation has been reported with increasing TDS concentration. Both are expected to decrease performance in membrane treatment systems (Winson et al). In a related observation, at higher TDS loadings the transmembrane pressure (TMP) increased over time across the MBR membrane leading to a higher fouling potential (Jang et al).

Table 3 – Wastewater Treatment System Biomass Characteristics Impacted by Variable TDS

Characteristic	Description	Impact of TDS Shock Loading
Extracellular Polymeric Substance (EPS)	Framework for holding together the internal constituents of cells, include high molecular weight polysaccharides, proteins, lipids and humic like substances.	Induces bursting, or lysing, of cells and the increase in EPS is a result of the release of intracellular contents that can increase fouling potential. Membrane fouling can be reversible or irreversible, requiring cleaning or replacement.
Zeta potential (mV)	Indirect measure of the electrical charge, typically negative, of the suspended floc in a wastewater treatment process	Reduces floc charge discouraging flocculation. The increased dispersion, or turbidity, may lead to blocking of membrane pores, such as those used in MBRs (Jang et al).
Particle size distribution (μm)	Size distribution of floc, or aggregated biomass	Reduces amount of filamentous bacteria that decreases stability of floc, and thus increases dispersion. Particle size of floc has been observed to be directly proportional to the step increase in TDS loading (mg/L) (Jang et al). Reduction in particle size and increase in density reported due to cell lysis (Winson et al).
Protozoa	Single-celled organisms that enhance flocculation and reduce turbidity by consuming microorganisms	Resistance to salinity shocks is limited. Protozoa unlikely to survive more than 24-hours after a NaCl shock higher than 40 g/L (Salvado et al).
Bio kinetic Coefficients	Substrate utilization rate; the rate at which microorganisms convert the influent nutrients.	General trend of lower substrate utilization rates with higher inorganic concentrations (Pophali et al).

Biomass Acclimation to TDS concentrations

The majority of microorganisms used in conventional biological treatment, such as activated sludge, are non-halophilic, and can tolerate TDS concentrations up to 10,000 ppm without acclimation (Panswad and Anan). Above this level, the outward flow of intracellular water due to osmotic stress increasingly impacts activity and ultimately leads to lysis. Also, above this level higher microorganisms such as protozoa tend to be absent. However, halophilic and halotolerant microorganisms possess the ability to manage osmotic stress. Halophilic microorganisms require a minimum salt level for survival. Halotolerant prefer a fresh water environment but can tolerate higher salt concentrations. Microorganisms can be classified by their optimal growth range in salt as shown in Table 4.

Different strategies may be appropriate for different changes in salt concentrations. Acclimation of microorganisms in conventional biotreatment systems to increasing salt concentration has been consistently reported up to 30,000 ppm, while an upper bound salt limit of 50,000 ppm has been observed for the acclimation strategy to work (Kargi and Dincer (1996) and Kargi (2002)). Acclimation duration is expected to be measured in over three days to weeks, while the acclimation achieved would also be non-permanent and lost when the salt environment changed. Thus non-conventional biological processes should be considered to treat wastewater that is expected to be subject to salt shocks of more than 30,000 to 50,000 ppm TDS. However, for salt concentrations over 30,000 ppm, addition of halophilic or halotolerant microorganisms should be considered as most biological functions would not be expected to remain functional through acclimation to higher concentrations (Winson et al). The important issue is to obtain the right type of microorganisms according to the salt range. In general, longer acclimation periods can be attributable to a transition in the microbial population (Intrasungkha et al), during which the dominant species of the population would change. Also, lower diversity of microorganisms at high TDS levels may reduce the stability of a biological process to environmental change (Salvado et al). Possible strategies to enhance the acclimation process include use of suspended support media for attached biofilm growth and operation of the biotreatment system at longer sludge retention times.

Biomass characteristics and biological operating conditions determine the efficiency of the biotreatment process. Therefore it is important to consider the effect of salt on the biomass characteristics and its relation to the operating conditions such as the organic loading rate, SRT and HRT, etc. If process conditions favor a particular range of TDS concentration and related class of microorganisms then a biotreatment system should be controlled to operate within this range, and should, to the extent possible, manage process variations to preserve the favored environment.

Table 4: Dominant microorganism populations vs. Observed biomass responses to shocks

Classifications of microorganisms per optimal growth range in salt (1)		Observed responses from un-acclimated Non-Halophilic biomass to shocks (2)
Classification	**Salt Range, ppm**	
Non-halophilic	<10,000	Tolerate without acclimation Above 8,000, possible increase in sludge yield (3).
Slightly halophilic	10,000 to 30,000	Acclimation period (days) expected
Moderately halophilic	30,000 to 150,000	Up to 50,000, prolonged Acclimation period (weeks) expected, usually benefits to add halophilic microorganisms. Above 50,000, consider adding halophilic microorganisms.
Extremely halophilic	>150,000	

Notes:

(1) Woolard and Irvine
(2) Winson et al
(3) Kincannon and Gaudy

Treatment System Design Considerations to Mitigate Variable TDS Levels

Achieving a reliable total system design requires an understanding of influent variability on both a constituent and hydraulic basis. In many process facilities, maintaining a consistent TDS concentration level in the influent to a wastewater treatment plant (WWTP) is not feasible. The WWTP needs to be designed to reliably manage the inevitable flow and concentration changes.

Since even an acclimated microbial population at high TDS tends to be sensitive to shock loads in TDS concentration, operating at a relatively constant salt concentration is important. In practical design terms, the need for equalization capacity is indicated as well as considering bio-treatment options that enhance operational stability.

Influent Wastewater Equalization

Equalization tanks and off-spec tanks for wastewater hold-up are typically used in two arrangements: in-line and off-line. All the flow passes through an in-line equalization tank. Only when streams are above a pre-determined value are they diverted into an off-spec tank (Gunaratnam and Smith).

The appropriately sized in-line equalization step provides mixing to dampen variations in concentration and retention time to dampen fluctuations in flow. This step also offers an opportunity for additional quality maintenance, such as temperature regulation and sampling and dosing, if required to prepare the wastewater for introduction to the biotreatment system.

Because effluent discharge limits are important considerations, including equalization capacity downstream would also provide additional flexibility in the event that the a system is shocked to the point of malfunction. An off-line effluent holding tank is often provided upstream of the outfall, when effluent quality must meet regulated quality standards. Effluent storage capacity should be at least equal to influent storage capacity (equalization capacity), so that that off-spec effluent could be recycled back to the equalization step for re- treatment. Some regulatory agencies provide guidance or recommendations on effluent hold-up capacity. Regardless of capacity, this additional point of manipulation in the system can provide sufficient capacity and opportunity for quality monitoring and sampling.

Biotreatment System Selection

The overall treatment system design must be robust, offering operational flexibility to manage upsets to normal influent quality and flow. However, biomass settling problems still have been commonly reported in saline environments, in part due to the higher density of salt water over freshwater, which results in higher buoyant forces. Membrane Bioreactors (MBRs) can provide a solution to settling problems, and hence effluent turdibity. Although reported biokinetic data at high TDS concentrations are usually case specific, increasing salt concentration results in slower biokentics. Consequently, operating at lower food to microorganism ratios and/or longer sludge retention times is recommended, which is an advantage to the use of MBRs.

When an influent TDS shock occurs, membrane based systems respond more successfully than conventional forms of biotreatment such as activated sludge (Jang et al). The ability to withstand shock loads more effectively is attributed to higher biomass concentrations in an MBR system, as well as the presence of a membrane. The membrane acts as a barrier to washout of solids and dead biomass, which can increase effluent organics.

Chloralkali Facility Case Study

Crucial elements for system design are determining an anchor point, and then an acceptable range of variability for TDS which can be effectively treated in the biotreatment step. The ability to treat high TDS wastewater is dependent on the appropriate quantity and type of microorganisms in a controlled environment. When the proper environment is achieved and maintained, the biotreatment system can successfully treat high TDS wastewater.

This case study involved retrofitting a bio-treatment system within an existing effluent treatment plant at a Chloralkali facility. The treatment specification for organics reduction was a two-tiered requirement. First, the effluent quality needed to allow for direct use as feed to a wastewater reuse unit that includes desalination. The second criteria was the applicable performance standard for direct discharge. Thus the effluent had to meet the more restrictive of these requirements. A portion of the available historical data for influent TDS is included in Figure 2.

Figure 2 - Historical Treatment Plan Influent TDS Levels

Stream Statistics		TDS mg/L
Average		29,000
Standard Deviation		11,000
90% Confidence Interval	Max	47,750
	Min	10,750
Median		30,000

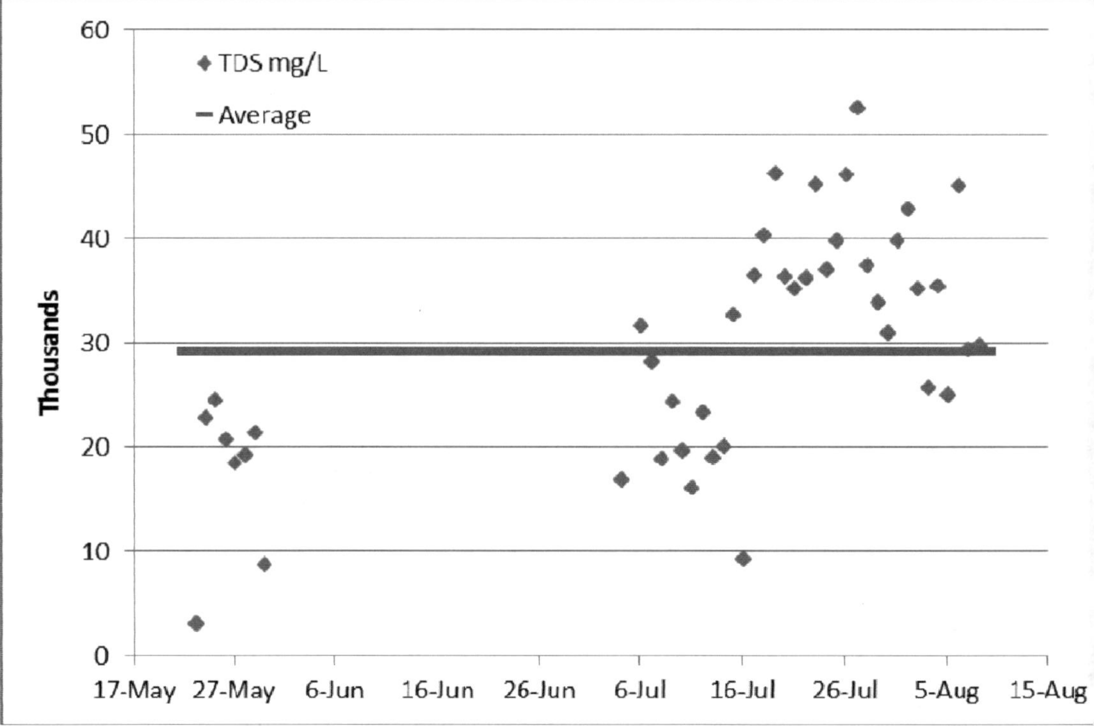

The average influent TDS, the anchor point for the biotreatment system, is 29,000 mg/L, a salt level at the high end of the slightly halophilic microorganism range displayed in Figure 3. Thus the constraint for TDS fluctuations in a licensor performance guarantee to a range of 20% per day from the average influent may prove too restrictive from a microbiological perspective. All other conditions being equal, a reduction in TDS concentration to 15,000 mg/L would not be problematic. However, justifying an increase of more than the 20% would be more difficult, as higher TDS levels in general lead to slower biokinetics and lower stability. In either case, balancing influent equalization and effluent holding tank capacity was to be used to maintain a feed TDS range to the biotreatment unit.

For this case study, the performance specification for in-line equalization capacity upstream of the biotreatment unit was to maintain TDS levels between 15,000 mg/l and 34,000 mg/l. The higher end value was 20% over the average influent value. However, the lower end value of 15,000 mg/l, shown in Figure 3 as an optional lower limit, was chosen to remain comfortably above the non-halophilic range to prevent shifting in microorganism populations. Because the historical data also showed 15,000 mg/l to be the lower end value of the 80% confidence interval, and thus only 10% of the time would the influent TDS value be less, a smaller equalization volume was required versus the 25% of the time that the TDS level would be below the 20% fluctuation from the average influent, or 23,000 mg/l. By not limiting the effluent from the equalization tank to a 20% TDS fluctuation of the average influent into the equalization tank, or a 23,000 to 34,000 mg/l range, the required working volume of the tank for TDS equalization purposes was able to be reduced.

Therefore with knowledge of process operations, an understanding of the physicochemistry and microbiology, and considerations for impact on biotreatment technologies, a shock tolerant treatment system can be designed.

Figure 3 -Analysis of Shock Load in Relation to Microorganism Optimal Growth Ranges in Salt

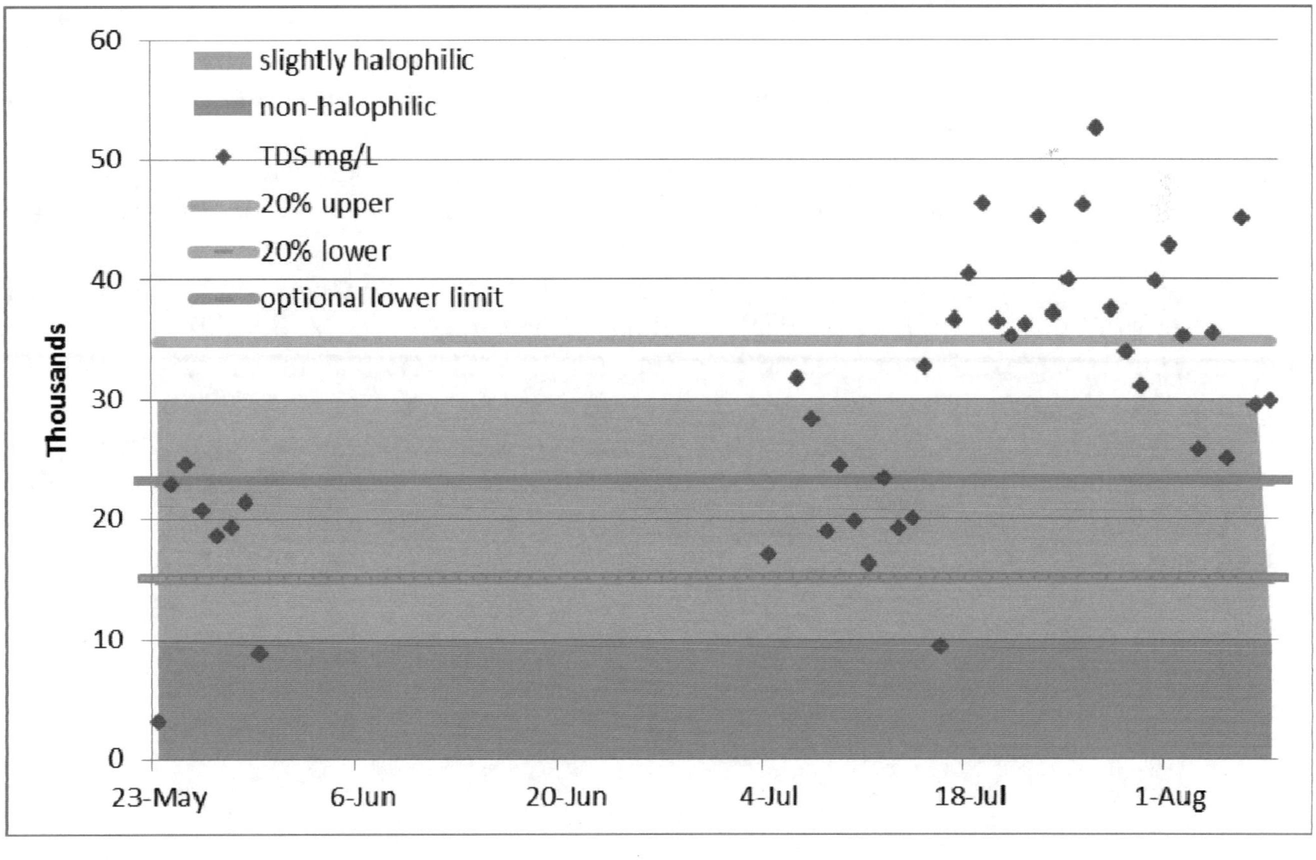

References

1. California Water Resources Control Board, Division of Water Quality, GAMA Program Groundwater Information Sheet – Salinity. Revised March 2010.
2. Diaz, MP, KG Boyd, SJW Grigson (July 20, 2002), "Biodegradation of Crude Oil across a Wide Range of Salinities by an Extremely Halotolerant Bacterial Consortium MPD-M, Immobilized onto Polypropylene Fibers", *Biotechnol. Bioeng.*, 79(2), 145-153.
3. Fair, GM; JC Geyer; DA Okun, Water and Wastewater Engineering, Volume 2 – Water Purification and Wastewater Treatment and Disposal, Wiley & Sons, Inc, 1968. Pg 30-2.
4. Gunaratnam, M and R Smith (2007), "Design of Buffering Capacity and Application to Design of Treatment Networks", *Trans IchemE, Part A, Chemical Engineering Research and Design* Vol 85 (A6), 852-862.
5. Intrasungkha, N, J Keller, LL Blackall (1999), "Biological nutrient removal efficiency in treatment of saline wastewater", *Water Science and Technology* 39 (6), 183-190.
6. Jang, D, Y Hwang, H Shin and W Lee (2013), "Effects of Salinity on the Characteristics of Biomass and Membrane Fouling in Membrane Bioreactors", *Bioresource Technology* 141, 50-56.
7. Kargi, F and AR Dincer (1996), "Effect of salt concentration on biological treatment of saline wastewater by fed-batch operation", *Enzyme and Microbial Technology*, 19, 529-537.
8. Kargi, F (2002), "Enhanced biological treatment of saline wastewater by using halophilic bacteria", *Biotechnology Letters*, 24, 1569-1572.
9. Kincannon, DF, AF Gaudy (1968), "Response of Biological Treatment Systems to Changes in Salt Concentrations", *Biotechnol, Bioeng.*, 10, 483-496.
10. Migliavacca, Michele, Manager of Business Development - Oil & Gas, Veolia Water Solutions & Technologies, Correspondence February 13, 2015.
11. Panswad, T, and C Anan (1999), "Impact of High Cloride Wastewater on an Anaerobic/Anoxic/Aerobic Process With and Without Inoculation of Chloride Acclimated Seeds", *Water Resources* Volume 33, No. 5, pp. 1165-1172.
12. Pophali, GR, SN Kaul, SMathur. (2003), "Influence of hydraulic shock loads and TDS on the performance of large-scale CETPs in India", *Water Research* 37, 353-361.
13. Salvado, H, M Mas, S Menendez, MP Gracia,. (2001), "Effects of Shock Loads of Salt on Protozoan Communities of Activated Sludge", *Acta Protozool*, 40 (3), 177-185.
14. Winson, CI, Lay, Y Liu, AG Fane (2010), "Impacts of salinity on the performance of high retention membrane bioreactors for water reclamation: A review", *Water Research* 44, 21-40.
15. Woolard, CR, RI Irvine (1995), "Treatment of hypersaline wastewater in the sequencing batch reactor", *Water Research* 29, 1159-1168.

401179 Simulation and Sensitivities of Membrane Networks for Desalination

Abstract:

Membranes are commonly used for the desalination and purification of either ground water or sea water. Based on the concentration of salt and the flow rate of the feed water networks of membranes should be constructed in order to reduce salt levels and to meet the required level of clean water production. However, the exact configuration and operating conditions required to meet these criteria will initially be unknown. For this reason a semi-rigorous model is built and implemented to simulate and predict the performance of the membranes. In this way it is possible to determine the numbers of membrane modules required and the inlet pressure required to achieve the desired separation.

A number of case studies are analysed to determine if alternative configurations or modified system parameters can enhance the system performance. In particular we test to see if different configurations of membrane module and energy recovery devices can be used while still meeting the constraints placed on water purity (salt rejection) and water recovery. This is important because optimizing membrane network configurations makes the process more energy efficient.

Acknowledgement

This work was supported by Korea Water Resources Corporation (K-Water), 2014.

15 AIChE
Spring Meeting, Austin, TX

ASSESSMENT OF DEFLUORIDATION OF DRINKING WATER USING NOVEL ADSORBENT NANO-CRYSTALLINE MAGNESIUM AMENDED HYDROXYAPATITE (M-HAP)

Dr. Suja George
Malaviya National Institute of Technology
Department of Chemical Engineering, Jaipur, Rajasthan, India
sgeorge.chem@mnit.ac.in

Poonam Mondal
Malaviya National Institute of Technology
Department of Chemical Engineering, Jaipur, Rajasthan, India
poonam.mbiotech@gmail.com

[Original work]

Prepared for Presentation at
American Institute of Chemical Engineers
2015 Spring Meeting
11th Global Congress on Process Safety
Austin, Texas
April 27-30, 2015

UNPUBLISHED

ASSESSMENT OF DEFLUORIDATION OF DRINKING WATER USING NOVEL ADSORBENT NANO-CRYSTALLINE MAGNESIUM AMENDED HYDROXYAPATITE (M-HAP)

Dr. Suja George
Malaviya National Institute of Technology
Chemical Engineering Department, Jaipur, Rajasthan, India
sgeorge.chem@mnit.ac.in

Poonam Mondal
Malaviya National Institute of Technology
Chemical Engineering Department, Jaipur, Rajasthan, India
poonam.mbiotech@gmail.com

Keywords: Fluoride, Adsorption, Nano-crystalline, Drinking water, Defluoridation.

Abstract

In the past few decades, fluoride contamination in drinking water has raised a great concern in the society as the earlier considered essential element fluoride turned into a major deleterious contaminant. Intake of fluoride into body in excessive amount (>1.5mg/L) causes adverse health effects such as dental and skeletal fluorosis and toxicity at molecular level. Endemic fluorosis is now known to be global in scope and occurring on all continents and millions of people are suffering the consequences. Water fluoridation has been banned in many countries including China, Austria, Belgium, Finland, Germany, Denmark, Norway, Sweden, the Netherlands, Hungary, and Japan. To counteract the effect of fluoride, a novel adsorbent was synthesized using chemical precipitation method and named as Magnesium amended hydroxyapatite (M-HAP) that can be used for defluoridation of fluoridated water. Characterization of M-HAP was performed via XRD, FTIR and TEM/EDX to analyze the structural composition, bonding patterns and micro-structural details respectively. The nano-crystalline structure (7-10nm) of M-HAP was confirmed through these studies. The defluoridation potential of M-HAP was investigated for defluoridation capacities through batch equilibrium experiments to gain insight of adsorption behavior, kinetics and mechanisms of adsorption of fluoride ion. The effect of various physicochemical parameters such as pH, adsorbent dose, initial fluoride concentration and the presence of interfering co-ions on adsorption of fluoride has been investigated. It was found that the adsorbent is very effective for defluoridation of drinking water with a maximum fluoride uptake capacity of 1.16mg/g achieved within 180minutes. Sorption process was found to be independent of pH and altered in the presence of other co-existing anions. The mechanism of adsorption is governed by Langmuir model and hence is monolayer in nature. The adsorption process follows pseudo-second-order and intraparticle diffusion kinetic models. Water quality parameters tested after treatment of fluoridated water with M-HAP indicates that the TDS, turbidity, hardness and total alkalinity are under permissible limit of standard drinking water regulations.

1. Introduction

Millions of people largely depend on groundwater with fluoride concentrations above the World Health Organization (WHO) guidelines (i.e. 1.5mg/L)[1]. High fluoride concentrations in groundwater are found in several countries such as USA, Africa, Asia, China, India, Ghana, Kenya, Tanzania , Sri Lanka , Australia and many other countries in different continents [2-3]. Consumption of fluoridated water more than the permissible limit causes fluorosis which is one of the most widespread endemic health problems associated with natural geochemistry. Excessive intake during developing stages of teeth leads to dental fluorosis and continued intake of fluoride for a long time may cause skeletal fluorosis. Even though it is an irreversible problem, it can be prevented by treating water using several defluoridation methods such as adsorption [4] membrane filtration [5] and electrocoagulation [6]. Among these techniques adsorption has emerged as the most economic and safe defluoridation technique [7]. Materials such as calcite, activated alumina, activated carbon have been used for the fluoride removal [8-10]. Activated alumina and other aluminum based materials are used as the conventional technology for defluoridation due to the high affinity of fluorine towards aluminium. The presence of residual alumina in treated water has attracted the investigators to replace the conventional technology with a novel alternative. Working in this direction it was found that Hydroxyapatite has been used for trapping and thus removing fluoride ions from solutions by several researchers [11-13]. Attributing to the fact that magnesium has a good affinity towards fluoride ions; a novel adsorbent magnesium amended hydroxyapatite is synthesized for defluoridation of drinking water. It is well known that when hydroxyl ions in Hydroxyapatite get replaced by fluoride ions, fluoroapatite is formed [14]. This mechanism is used for removing fluoride from aqueous solution.

2. Materials and Methods

2.1 Synthesis of M-HAP

Magnesium amended HAP (M-HAP) was synthesized in the laboratory through chemical precipitation technique by drop-wise addition of potassium dihydrogen phosphate solution(0.32M) into a solution consisting of magnesium chloride and calcium nitrate with Mg/Ca molar ratio of 1.16. The pH of the suspension was adjusted to 10 for 5-6 hours while constantly stirring the solution. The resulting suspension was cooled and kept overnight at room temperature and a white gel like precipitate was obtained. This precipitate was further dried at a temperature of 105 ± 2°C and used after crushing to powder form.

2.2 Fluoride adsorption experiments

Adsorption experiments in batch mode were conducted in the laboratory at a temperature of 25 ± 0.5°C. After the equilibrium was reached, the liquid samples were filtered through Whatman 42 filter paper and analyzed for residual fluoride using an ion selective electrode (Orion Versa Star 93 meter). Defluoridation studies were carried out under various experimental conditions, such as pH, contact time, initial fluoride concentrations, adsorbent dose and the presence of competing ions. All experiments were done in triplicates to reduce experimental error. The data obtained from the experiments were fitted with different models to identify the adsorption mechanism.

3. Results and Discussion

3.1 Characterization of M-HAP

The characterization of M-HAP was done by means of FTIR, XRD, and TEM with EDX. Study of FTIR spectra of M-HAP in Fig 1 illustrates OH^- band at 3571.8 cm^{-1} and PO_4^{3-} bands at 601.8 cm^{-1},566.78cm^{-1} and 1041.7 cm^{-1} associated with hydroxyapatite. The broad band for adsorbed water (3000–3500 cm^{-1}) is indication of water absorption due to the high specific surface area that precipitated powders usually possess. XRD diffractogram in Fig 2 has very sharp and narrow peaks which clearly depicts crystalline nature of the adsorbent synthesized. The pattern matches to JCPDS card number 00-09-0432 for Hydroxyapatite very well. A peak for magnesium oxide is also observed at 2 theta angle of 20.94 which represents peak for magnesium oxide (JCPDS: 00-027-0759). Therefore, it can be inferred that magnesium is successfully incorporated in hydroxyapatite structure. TEM analysis of the adsorbent depicts that the particles are aggregated with an average particle size of 7-10nm(Fig 3).EDX analyses were performed to determine the elemental constituents of M-HAP which shows calcium, phosphorus, magnesium and oxygen(Fig 4).

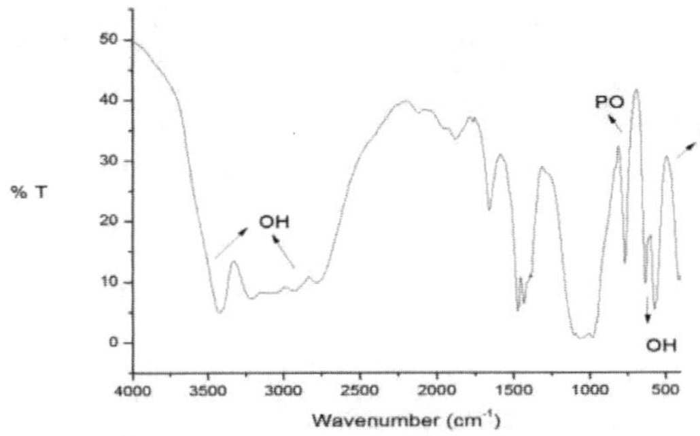

Fig 1: FTIR spectra of M-HAP

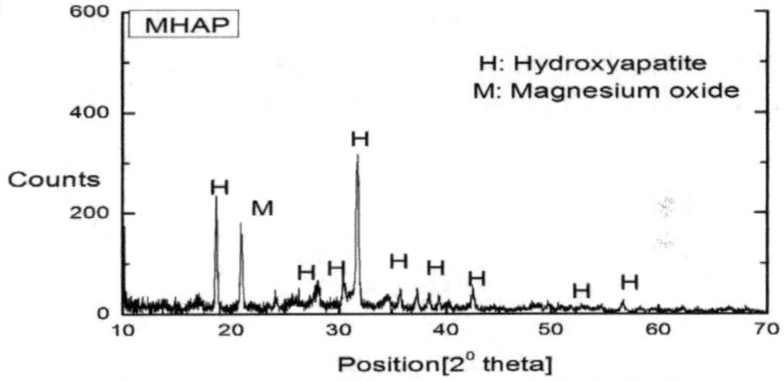

Fig 2: XRD diffractogram of M-HAP

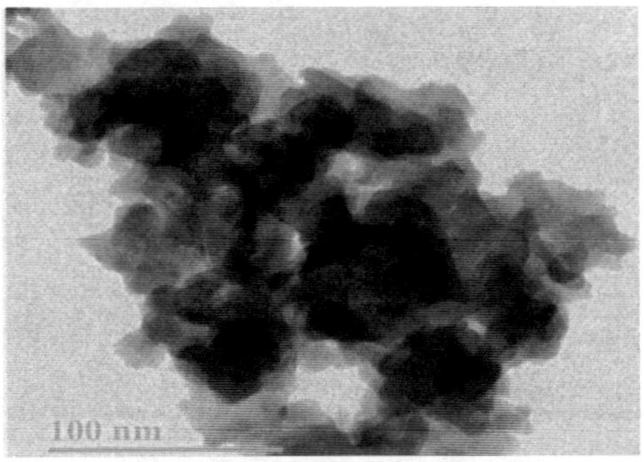

Fig 3: TEM image of M-HAP

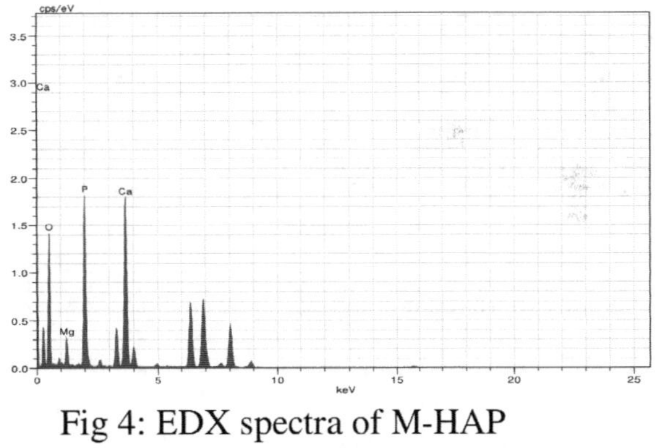

Fig 4: EDX spectra of M-HAP

3.2 Adsorption experiments

3.2.1 Effect of adsorbent dose

The effect of adsorbent dose on the adsorption of fluoride is presented in Fig 5. It was observed that the percentage removal of fluoride increased from 41% to 95% with an increase in M-HAP dose from 2 to 17 g/L. However, after a dosage of 10 g/L, the amount removed per unit of adsorbent remained almost constant. Greater concentration of the adsorbent (>10 g/L) will increase cost of the process, without introducing a significant benefit in the amount of fluoride removed. Therefore, the dose 10 g/L was considered sufficient to reduce the amount of fluoride to the WHO drinking water guideline and used in further experiments.

3.2.2 Effect of contact time

The sorption of fluoride on M-HAP was investigated as a function of contact time at a concentration of 10mg/L. More than 85% removal of fluoride was attained within first 90 min of contact time and equilibrium was achieved in 180 min (Fig 6). The greater sorption rate at initial period of time is due to increased number of vacant sites as a result of which there is an increased concentration gradient between adsorbate and adsorbent surface. However, as time passes this concentration is reduced and eventually reaches to an equilibrium point. The data obtained from this experiment is used to assess the kinetics of the adsorption process.

3.2.3 Effect of pH

The pH controls the adsorption at the water-adsorbent interfaces. The equilibrium sorption of fluoride (with initial fluoride concentration of 10 mg/L) was investigated over a pH range of 3–9 using 0.1 M NaOH and HCl solutions. It can be seen (Fig 7) that the sorption of fluoride on M-HAP is pH independent which is a great advantage for practical usage of the adsorbent.

3.2.4 Effect of initial fluoride concentration

With increase in initial fluoride concentration, the percentage removal of fluoride decreases. The initial fluoride concentration was increased from 2 to 20 mg/L and the removal efficiency reached a maximum value of 1.113mg/g as shown in Fig 8.

3.2.5 Effect of competing ions

Depending on the geographical location, various ions may be present in groundwater which can compete with the uptake of fluoride ions by M-HAP. The effect of co-existing anions (Cl^-,NO_3^{2-},SO_4^{2-},HCO_3^{2-} and PO_4^{3-}) on fluoride adsorption capacity was studied at constant fluoride concentration of 10 mg/L and adsorbent dose of 10 g/L, by varying anion concentration from 100 to 300 mg/L. The orders of interference by these ions are as follows: Chloride < Sulfate <Nitrate <Phosphate< Bicarbonate (Fig 9).

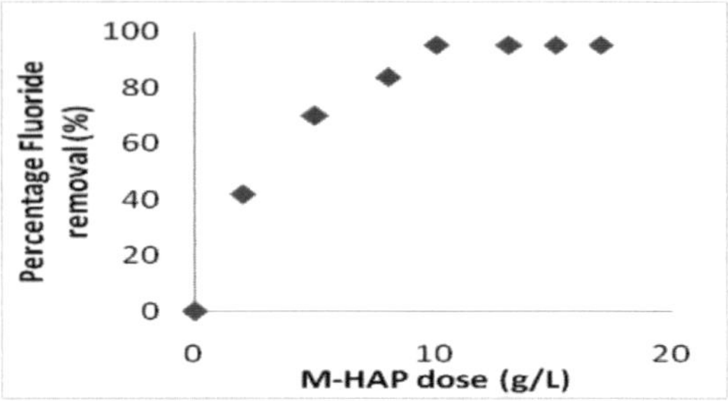

Fig 5: Effect of adsorbent dose

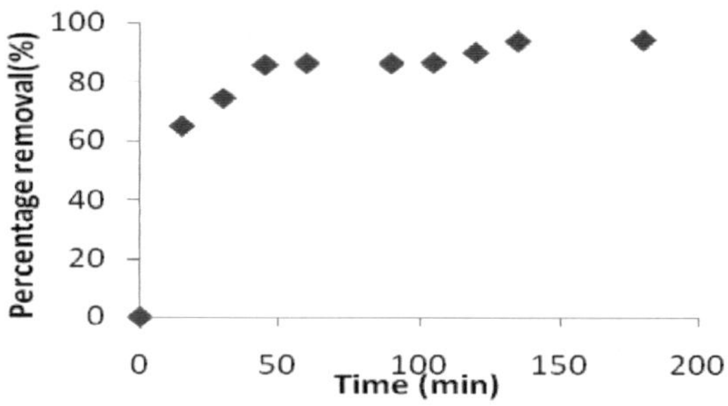

Fig 6: Effect of contact time

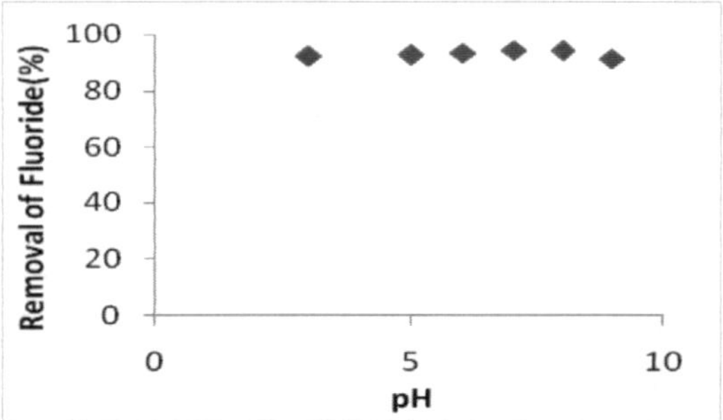

Fig 7: Effect of pH

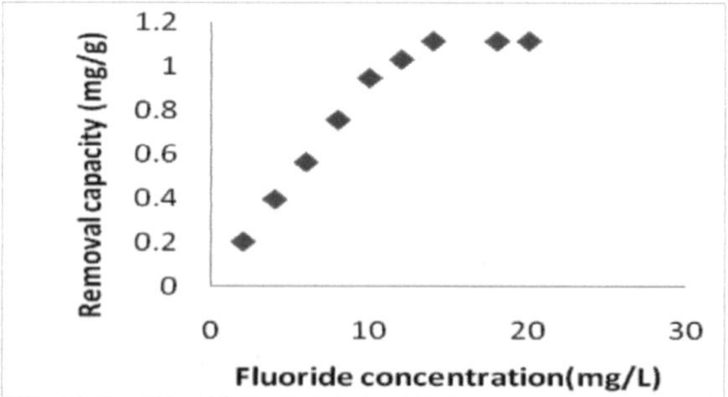

Fig 8: Effect of initial fluoride concentration

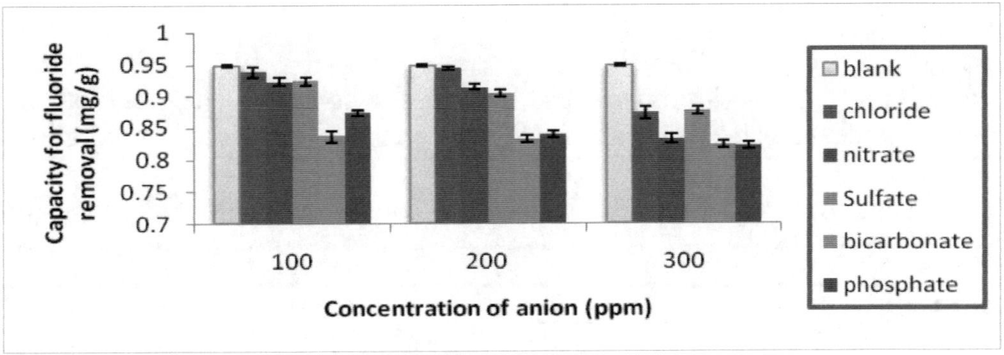

Fig 9: Effect of competing ions

3.3. Equilibrium studies

The correlation of equilibrium adsorption data by either theoretical or empirical equations is important in the design and operation of adsorption systems. The present study employed the Langmuir and Freundlich models to describe the equilibrium adsorption. The expression of the Langmuir model is given in Table 1 where q_e is the amount of fluoride adsorbed per gram of M-HAP (mg/g); C_e denotes the equilibrium concentration of fluoride in solution (mg/L); K_L represents the Langmuir constant (L/mg) that relates to the affinity of binding sites and q_m is the theoretical saturation capacity of the monolayer (mg/g). Furthermore, the effect of the isotherm shape is considered with a view to predict whether an adsorption system is favorable or unfavorable. Another important parameter, R_L, called the separation factor is also evaluated in this study, it is determined from the relation:

$$R_L = \frac{1}{1 + K_L C_0}$$

Where K_L is the Langmuir constant (L/mg) and C_0 is the initial fluoride concentration (10 mg/L). The value of R_L is less than 1 which indicates favorable adsorption.

The Freundlich model is an empirical equation that assumes heterogeneous adsorption due to the diversity of adsorption sites. The equation is given in table 1 where q_e is the equilibrium fluoride concentration on M-HAP (mg/g); C_e the equilibrium fluoride concentration in solution (mg/l); K_F and n are the Freundlich constants, which represent the adsorption capacity and the adsorption strength, respectively. The magnitude of 1/n quantifies the favorability of adsorption and the degree of heterogeneity of the M-HAP surface.

3.4 Kinetics studies

The kinetics of fluoride sorption on nano-crystalline M-HAP was studied using pseudo-first-order and pseudo-second-order kinetic models to identify the dynamics of the solute adsorption process. The best-fit model, i.e. pseudo-second-order model was chosen based on the match between experimental (q_e (exp)) and theoretical (q_e (cal)) values and the linear correlation coefficient (R^2).It was observed that the fluoride uptake increases with the drift of time. The adsorption of fluoride was rapid at first after which the rate slowed down as equilibrium approached. The results obtained from the experiments were used to study the rate-limiting step.

The values obtained by pseudo second order model were found to be in good accord with experimental data and can be used for explaining the fluoride sorption on M-HAP (Table 2).

While studying the kinetics intraparticle diffusion was also considered which is described by the equation:

$$q_t = k_i t^{1/2}$$

where k_i (mg $g^{-1}min^{-1/2}$) is the intra-particle diffusion rate constant. The slope of the plot of q_t against $t^{0.5}$ will provide the value of intra particle rate constant which is 0.114 mg $g^{-1}min^{-1/2}$.

Table 1: Langmuir and Freundlich isotherm constants for fluoride adsorption

Isotherm	Equation	Parameters	Value
Langmuir	$\frac{1}{q_e} = \frac{1}{q_m} + \frac{1}{q_m b C_e}$	q_m(mg/g)	1.16
		b(L/mg)	5.937
		R^2	0.991
		R_L	0.011
Freundlich	$lnq_e = lnK_f + n\ lnC_e$	K_f [mg/g(L/mg)$^{1/n}$]	0.857
		n	4.149
		R^2	0.988

Table 2: Kinetic models with their parameters for fluoride sorption

Kinetic model	Equation	Parameters	Value
Pseudo first order	$\log(q_e - q_t) = \log q_e - \frac{k_1 t}{2.303}$	q_e(mg/g)	0.406
		k_2 (min^{-1})	0.023
		R^2	0.705
Pseudo second order	$\frac{t}{q_t} = \frac{1}{k_2 q_e^2} + \frac{1}{q_e t}$	q_e(mg/g)	0.981
		k_2 (g mg^{-1} min^{-1})	0.113
		R^2	0.997

3.5 ***Physico-chemical properties of treated water***

The quality of water treated with M-HAP adsorbent was tested and the TDS, Total alkalinity and Total hardness was 148 mg/L, 75 mg/L and 240 mg/L respectively [15].No turbidity was observed in the sample solutions. All the physicochemical properties were under permissible limit [1]

Conclusion

The necessity to find a defluoridation media for safe and easy use at both household and community level is long desired. While assessing the capability of M-HAP adsorbent for defluoridation of drinking water, it was found that the adsorbent works well in a broad pH range with a maximum uptake capacity of 1.16mg/g. This capacity was observed to be affected majorly by presence of bicarbonate and phosphate anions and slightly by nitrate, sulfate and chloride anions. Equilibrium was reached in 180minutes and equilibrium studies showed that the adsorption is monolayer in nature since Langmuir isotherm well fitted in the data. Moreover, kinetic studies revealed that the adsorption follows pseudo second order reaction and intraparticle diffusion also governs the mechanism. The quality of treated water is the added advantage of using M-HAP as a defluoridating agent and therefore it can further be designed and developed for domestic defluoridation units.

References

[1] WHO. Guidelines for drinking-water quality.Volume 1 recommendations, 3rd edn. World Health Organization,Geneva. 2004

[2] Susheela, A. K., "Fluorosis management programme in India, *Current science*,VOLUME 77, NO 10, 1999.

[3] Jagtap, S., Yenkie, M. K., Labhsetwar, N., &Rayalu, S., "Fluoride in drinking water and defluoridation of water", *Chemical reviews*, VOLUME 112, NO 4, 2012.

[4] Mondal P., George S., "A review on adsorbents used for defluoridation of drinking water", *Reviews in Environmental Science and Bio/Technology*, 2014 DOI 10.1007/s11157-014-9356-0

[5] P. Sehn, "Fluoride removal with extra low energy reverse osmosis membranes: three years of large scale field experience in Finland", *Desalination*, VOLUME 223, NO 1,2007

[6] Essadki, A. H., Gourich, B., Vial, C., Delmas, H., &Bennajah, M., " Defluoridation of drinking water by electrocoagulation/electroflotation in a stirred tank reactor with a comparative performance to an external-loop airlift reactor" *Journal of hazardous materials*, VOLUME 168,NO 2,2009.

[7] Fan, X., Parker, D.J., & Smith, M.D. "Adsorption kinetics of fluoride on low cost materials", *Water Research*, VOLUME 37,NO 20, 2003

[8] Mondal P., George S., Mehta D., "Use of Calcite for Defluoridation of Drinking Water in Acidic medium", *Research Journal of Chemical Sciences*, Volume 4, NO. 6, 2014

[9] George S., Pandit P., Gupta A.B., "Residual aluminium in water defluoridated using activated alumina adsorption–Modeling and simulation studies", *Water Research*, Volume 44, NO.10,2010

[10] Karthikeyan, G., &Ilango, S. S. "Fluoride sorption using *Morringa Indica*-based activated carbon". *Iranian Journal of Environmental Health Science & Engineering*, VOLUME 4, NO 1, 2007.

[11] Gao, S., Sun, R., Wei, Z., Zhao, H., Li, H., & Hu, F., "Size-dependent defluoridation properties of synthetic hydroxyapatite",*Journal of Fluorine Chemistry*, VOLUME 130, NO 6, 2009.

[12] Sundaram, C.S., Viswanathan, N., and Meenakshi, S. "Fluoride sorption by nanohydroxyapatite /chitin composite". *Journal of Hazardous Materials*, VOLUME 172, NO 1, 2009.

[13] Liang W, Zhan L, Piao L, Russel C. "Fluoride removal performance of glass derived hydroxyapatite", *Material Research Bulletin*, VOLUME 46, NO 2, 2011.

[14] Cao, W., & Hench, L. L. Bioactive materials. *Ceramics international*, VOLUME 22, NO 6, 1996

[15] APHA. Standard Methods for the Examination of Water and Wastewater.19th Edition. American Public Health Association, Washington, DC. 1995

AUTHOR INDEX